Science, Technology and Industry in the Ottoman World

DE DIVERSIS ARTIBUS

COLLECTION DE TRAVAUX
DE L'ACADÉMIE INTERNATIONALE
D'HISTOIRE DES SCIENCES

COLLECTION OF STUDIES
FROM THE INTERNATIONAL ACADEMY
OF THE HISTORY OF SCIENCE

DIRECTION
EDITORS

EMMANUEL POULLE   ROBERT HALLEUX

TOME 46 (N.S. 9)

BREPOLS

PROCEEDINGS OF THE XXth INTERNATIONAL CONGRESS OF HISTORY OF SCIENCE (Liège, 20-26 July 1997)

VOLUME VI

SCIENCE, TECHNOLOGY AND INDUSTRY IN THE OTTOMAN WORLD

Edited by

Ekmeleddin IHSANOGLU, Ahmed DJEBBAR, and Feza GÜNERGUN

BREPOLS

The XXth International Congress of History of Science was organized by the Belgian National Committee for Logic, History and Philosophy of Science with the support of :

ICSU
Ministère de la Politique scientifique
Académie Royale de Belgique
Koninklijke Academie van België
FNRS
FWO
Communauté française de Belgique
Région Wallonne
Service des Affaires culturelles de la Ville de Liège
Service de l'Enseignement de la Ville de Liège
Université de Liège
Comité Sluse asbl
Fédération du Tourisme de la Province de Liège
Collège Saint-Louis
Institut d'Enseignement supérieur "Les Rivageois"

Academic Press
Agora-Béranger
APRIL
Banque Nationale de Belgique
Carlson Wagonlit Travel - Incentive Travel House
Chambre de Commerce et d'Industrie de la Ville de Liège
Club liégeois des Exportateurs
Cockerill Sambre Group
Crédit Communal
Derouaux Ordina sprl
Disteel Cold s.a.
Etilux s.a.
Fabrimétal Liège - Luxembourg
Generale Bank n.v. - Générale de Banque s.a.
Interbrew
L'Espérance Commerciale
Maison de la Métallurgie et de l'Industrie de Liège
Office des Produits wallons
Peeters
Peket dè Houyeu
Petrofina
Rescolié
Sabena
SNCB
Société chimique Prayon Rupel
SPE Zone Sud
TEC Liège - Verviers
Vulcain Industries

D/2000/0095/49
ISBN 2-503-51095-7
Printed in the E.U. on acid-free paper

TABLE OF CONTENTS

Foreword 7
Ekmeleddin IHSANOGLU, Ahmed DJEBBAR, Feza GÜNERGUN

Ottoman Science : The Last Episode in Islamic Scientific Tradition and the Beginning of European Scientific Tradition 11
Ekmeleddin IHSANOGLU

Les activités mathématiques au Maghreb à l'Epoque Ottomane (XVIe-XIXe siècles) 49
Ahmed DJEBBAR

Les différents systèmes de numérotation au Maghreb à l'Epoque Ottomane : L'exemple des chiffres rūmī 67
Youcef GUERGOUR

Introduction à l'étude de l'influence d'Ibn al-Bannā sur les mathématiques en Egypte à l'Epoque Ottomane 75
Mohamed ABALLAGH

The Development in the Attitude of the Ottoman State towards Science and Education and the Establishment of the Engineering Schools (*Mühendishanes*) 81
Mustafa KAÇAR

Scientific Motivation for and mood from the Experience of the Egyptian Expedition 91
Jean DHOMBRES

L'expérience préalable de l'Empire ottoman dans la Commission des sciences et arts de l'expédition de Bonaparte en Egypte (1798-1801) 101
Patrice BRET

The Introduction and Reception of Modern Science and Technology in " Ottoman " Egypt in the Nineteenth Century 115
A.H. Helmy MOHAMMAD

An Ottoman Professor of Botany : Salih Efendi (1817-1895) and his Contribution to Botanical Education in Turkey 127
Feza GÜNERGUN

The Imperial Ottoman Izmir-to-Aydin Railway : The British Experimental Line in Asia Minor.......... 139
Yakup BEKTAS

Contributors.......... 153

Foreword

Ekmeleddin IHSANOGLU - Ahmed DJEBBAR - Feza GÜNERGUN

In recent years the international history of science congresses organized by the IUHPS have devoted increasing attention to the history of science and technology in Islamic civilization. This attention was institutionalized at the XVIIIth International Congress in Hamburg in 1989 with the establishment of an International Commission on Science and Technology in Islamic Civilization within the IUHPS. Henceforth researchers working in this area could organize and coordinate their studies within the framework of this commission. At the same time, studies on the cultural sub-groups of Islamic civilization started to attract more scholarship than ever before, as did historical epochs which had not previously drawn the attention of historians of science.

As part of this trend, the scientific activities of the Ottoman Empire (1299-1923) began to receive more attention from historians of science, for two reasons. First, the Ottoman period was an under-studied episode in the Islamic scientific tradition. Second, the Ottoman world was the first non-Western intellectual culture to have contact with emerging Western science. The symposia on Ottoman science organized by IRCICA (International Research Center for Islamic History, Art and Culture) and TBTK (Turkish Society for History of Science) in the last two decades were influential in fostering and reviving studies on Ottoman science. The proceedings of these symposia, and especially the series " Ottoman Scientific Literature ", published by IRCICA from 1990 on, were influential in this revival. The first four volumes of this series, which cover the important Ottoman works on astronomy and mathematics, bring the wealth of Ottoman scientific literature to the attention of historians of science. The book reviews published for these four volumes emphasized two features of the Ottoman period : first, the encounter of Islamic and Western traditions which characterizes this period, and second, the small amount of existing scholarship on Ottoman science.

Thus, the symposium " Science, Technology and Industry in the Ottoman World " was organized by Ekmeleddin Ihsanoglu and Ahmed Djebbar at the xxth International Congress of History of Science (Liège, 20-26 July 1997) in response to the need for more scholarship on Ottoman science. This symposium was also an important step in defining the place of the history of Ottoman science within the broader history of science. The present volume contains the papers presented at the Liège symposium, and seeks to understand the scientific activities carried out in various geographical areas of the Ottoman Empire, including Turkey, the Maghreb, and Egypt. The plenary lecture given by E. Ihsanoglu at the Liège Congress is included. This lecture gives an overall picture of scientific activities throughout the 600-year history of the Ottoman Empire, thus bringing to light a new aspect of Ottoman history — the history of Ottoman science.

Three papers examine the history of mathematics in the Maghreb. A. Djebbar surveys the mathematical activities carried out in the Maghreb in both pre-Ottoman (12th-15th centuries) and Ottoman (16th-19th centuries) times. Relying on his studies on Maghrebian mathematical literature, he demonstrates that continuity was the principal characteristic of Maghrebian mathematics. For example, the phenomenon of versification, which flourished in the 16th century, had its origins in the 12th century. Youcef Guergour examines the 14th century Maghrebian mathematician Ibn al-Banna's work *al-Iqtidab min al'amal bi r-rumi fi l-hisab*, which explains how to perform mathematical calculations with *rumi* numerals. This and other such works claim that *rumi* numerals were widely used in the Maghreb in commercial transactions and in administrative services starting from 12th century. Mohamed Aballagh shows that the commentary on Ibn al-Banna's work *al-Talkhis*, by the Egyptian mathematician Ibn al-Majdi, contributed to the introduction of Maghrebian mathematics into pre-Ottoman Egypt. Thus a synthesis of Eastern and Western Islamic mathematics occurred in Egypt in the 15th century.

Mustafa Kaçar's paper sheds light on the reforms in military technical training in 18th century Ottoman Turkey. The Corps of Salaried Bombardiers (est. 1735) and the Engineering School (est. 1775), which were created on the initiatives of Ottoman administrators, and the technical knowledge of European experts were both influential in the introduction of modern science and technology to the Ottoman State.

Three papers focus on late 18th and early 19th century Ottoman Egypt. In his paper on the Egyptian expedition, Jean Dhombres tries to identify the intellectual motivations which led French scientists and engineers to explore Egypt. He describes these French intellectuals as " a small and often disoriented community ", but concludes that in their motivations for going to Egypt " something purely scientific was at stake ". He argues that the indifference of Egyptians to modernity provides an explanation for the failure of French colonialism in Egypt.

Patrice Bret looks at the engineers and technicians who participated in Bonaparte's expedition to Egypt in 1798, especially those with previous experience in Ottoman Turkey. He discusses the ways in which their experience was put to use (or misuse) in Egypt. He identifies and compares their achievements in Ottoman Turkey and Egypt respectively, and shows that French experts did not cooperate with the Egyptians in the improvement of the country, as they had with the Turks.

A.H. Helmy Mohammed surveys the efforts of Mohammed Ali Pasha (r. 1805-1849) to introduce modern science and technology to Egypt by founding a Medical School in 1827, sending students to Europe for training (1813), and starting a movement to translate European scientific books into Arabic. Two monthly magazines, *al-Moqtataf* and *al-Hilal*, were also key in introducing new scientific concepts to Egyptian society. In his conclusion, the author compares this 19th century transfer of scientific knowledge with the one that occurred in the Caliph al-Mamun's time (9th century), when the scientific knowledge of Antiquity was transmitted to a newly emerging Islamic civilization.

Yakup Bektas investigates the first railway line in Asiatic Turkey, the Izmir -Aydin railway (1857-1866), constructed by the British as part of their imperial expansion to the East. This line, together with the Izmir-Kasaba railway (1863-1866), enjoyed a fair measure of commercial success. This success further stimulated enthusiasm within the Ottoman government for the expansion of rail transportation, as well as wider public interest. Although the Ottoman railways were initially financed, constructed, and operated by European entrepreneurs, the Hamidiye-Hicaz Railway (1900-1908), which was financed from Muslim sources and constructed by Ottoman engineers, stands as a remarkable early 20th century Ottoman achievement.

The modern medical schools established in Istanbul in the first half of the 19th century introduced the teaching of natural sciences into Turkey. Thus, botany was first taught by physicians, among them Salih Efendi, the chief physician to the Sultan. In her paper, Feza Günergun examines the two editions of *Ilm-i Hayvanat ve Nebatat* (Istanbul, 1863, 1870), the botany book Salih Efendi translated from French, and brings to light the original work from which it was translated.

It is our hope that these papers will generate further scholarly interest in the history of science and technology in the Ottoman period. In offering this collection, we would like to acknowledge Professor Robert Halleux and his team in Liège for their kind help in realizing this symposium as part of the IUHPS Congress, and for their generous support in the publishing of the proceedings.

OTTOMAN SCIENCE : THE LAST EPISODE IN ISLAMIC SCIENTIFIC TRADITION AND THE BEGINNING OF EUROPEAN SCIENTIFIC TRADITION

Ekmeleddin IHSANOGLU

Upon consideration of the scientific activities carried on throughout the six-century history of the Ottoman Empire (1299-1923), it may be argued that the history of Ottoman science witnessed several distinctive trends. Though the historical evolution of Ottoman science shared many features common to the history of scientific endeavor in other parts of the Muslim world (beyond the boundaries of the Ottoman Empire), there were also some important differences. In the process of this historical evolution, the Ottomans were considered " pioneers " in some areas of scientific endeavour.

Inspired by the medieval Islamic scientific tradition at the beginning of its history, Ottoman science quickly became influential in the old scientific and cultural centers of the Islamic world. However, when in the seventeenth century the Ottomans were introduced to European science, they became instrumental in spreading this new scientific tradition through the Muslim lands. Thus, the Ottomans, who in their day were the representative of the whole Islamic world, were able to combine Islamic scientific tradition with the newly emerging Western science. Around the turn of the nineteenth century, however, the Ottomans opted for the exclusive practice of European science over Islamic, and the Islamic scientific tradition gradually faded away as a new tradition emerged in accordance with modern Western scientific norms. This paper aims to draw a general picture of the history of Ottoman science, and show the role it played in bringing together two different civilizations. It also attempts to survey the scientific activities conducted by the Muslims that formed the majority of the population in the Ottoman Empire.

Scientific activities were, of course, also undertaken among the non-Muslim population of the Ottoman Empire, which was comprised of Jews and various Christian subjects, including Greeks, Armenians, Bulgarians, Serbs, Romanians, and Hungarians. The scholarly and scientific activities of these people, which were recorded in languages other than Turkish, Arabic, or Persian have not been included in this paper. The main reason for this exclusion is that studies on the scientific activities of non-Muslim Ottoman subjects are as yet insufficient in number to allow us to offer a general overview of this aspect of the history of Ottoman science.

The first part of this survey deals with the Ottoman absorption of classical Islamic scientific tradition. This tradition came to the Ottomans by way of a pre-Ottoman Islamic scientific legacy — most importantly, the scientific legacy of the Seljuk Turks, who had ruled Anatolia and the Middle-East before the Ottomans. The second part will analyze modern Western scientific tradition, which was introduced to and developed within the Ottoman world as a result of the Ottomans' close relations with Europe in later centuries.

PRE-OTTOMAN ISLAMIC SCIENTIFIC TRADITION

In the first centuries of the history of Islam and the first stages of the political and legal formation of Islamic civilization, teaching activities and education were conducted mainly in mosques. Translations made from Greek and Indian sources, which sought to introduce those sciences called *Ulumu'l-evail* (the sciences of the ancients, including philosophy, mathematics, medicine, astronomy, physics, chemistry, and so on) led to the creation of new institutions to house these new scientific activities. An early example is the *Beytü'l-hikme* which was founded in Baghdad. Institutions of this type, later known as *Darü'l-hikme* and *Darü'l-ilm*, dealt mainly with non-religious sciences and were founded under the patronage of the sovereigns.

In the eleventh century A.D., a new educational institution called the *medrese* was established in the Eastern part of the Islamic world, in Iraq, Iran, and Khorasan. The main focus of the formal education offered in the *medrese* was Islamic jurisprudence and related religious sciences. The *medreses* were supported by pious endowments (*waqfs*) that were established by individuals for the sake of God. The State did not directly interfere with education or curriculum in the *medrese*. The content of the curriculum and methods of instruction were based on the historical evolution of the teaching profession and the dynamic relationship between the various religious and intellectual trends.

In the Great Seljuk period, the *medrese* (with its traditionally adjoining dervish monastery) became the most popular form educational institution. Due to the efforts of Nizamü'l-Mülk (1015-1092), the Grand Vizier of Sultan Alparslan (and later of his son, Sultan Melikshah,) *medreses* became widespread. Nizamü'l-Mülk founded a " Nizamiye Medrese " in every prominent city in the Seljuk Empire, the most famous among these being the Baghdad Nizamiye Medrese, founded in 1065.

According to G. Makdisi, the institutional education offered within the medieval Islamic lands was concerned solely with " religious " sciences. The legal structure of the *medrese* system was developed by Islamic jurists (*fuqaha*). Consequently, *fiqh* (Islamic law) studies were given priority in the *medrese* system, and rational sciences were not included in the curriculum. Higher officials, who would go on to be appointed to religious, juridical, and political posts within the Empire, were educated and trained according to the *sunni* doctrine in these *medreses*. Before long, the establishment of *medreses* became widely popular in the *sunni* community, and, as a result of the State's expanding organization, these *medreses* clustered in specific centers.

Foundation of new *medreses* continued at full throttle in the smaller Muslim states that eventually branched off of the Great Seljuk Empire. During the reign of Nureddin Zengi (d. 1174) and Selahaddin Eyyubi (d. 1193), members of the royal family and prominent statesmen began to establish *medreses*, hospitals, and dervish monasteries in different cities. In this period, the *medreses* even spread to small towns, where many charitable and educational institutions were founded.

Islamic culture and science enjoyed its most glorious period under the Seljuks, and during the subsequent period when smaller states budded off of crumbling Seljuk Empire. The writings of many famous scholars, works of art, buildings, and other architectural monuments that have come down to us attest to the glory of this period, and preserve the memory of its magnificence and power even into the present day. During the era of the Seljuks and their followers, cities such as Baghdad, Merv, Isfahan, Nishapur, Mosul, Damascus, Cairo, Aleppo, Amid (Diyarbakir), Konya, Kayseri, and Malatya all became a scholarly centers thanks to the establishment in those places of cultural, educational, and medical institutions.

After Cengiz Khan's (d. 1227) invasion, cultural and scientific activities in the Islamic world came to a standstill. However, learning was to regain momentum a century later during the Ilkhanid period, specifically during the reigns of Hulagu (d. 1265) and Gazan Khan (d. 1304), and later during the Muzafferid period. The observatory founded in the town of Meraga in Azerbaijan and the attached school of mathematics and astronomy, together with the cultural activities in the adjacent regions of Iran and Transoxania were behind this revival of learning. Scientific and scholarly activities were greatly influenced by this new tradition. In Gazan Khan's period Tabriz replaced Meraga as the leading center of learning. *Medreses* in the cities of Shiraz, Kirman, and Yazd also made significant advances during the Muzafferid period.

The second great Central Asian emperor, Timur (d. 1405), reigned over a vast realm extending from the borders of China to the Ukraine, and occupied India, Iran, Iraq, Azerbaijan, Turkistan, southern Russia, Anatolia, Syria, and Egypt between 1370 and 1405. During his reign, Timur patronized *medreses*, libraries, and other institutions of learning, as well as funded the development of cultural institutions in Turkistan, Khorasan, and Transoxania. Indeed, his patronage of such institutions of learning was extensive by comparison with other contemporary rulers. Consequently, during the reigns of Timur and his son Shahruh (d. 1447), Samarkand and Herat became famous centers of learning and culture, attracting scholars, artists, and students from all over the Islamic world.

Timur conquered the Ottoman lands briefly ; but after his death, the peace established between the Ottomans and the Timurids during the 43-year reign of Timur's son Shahruh was instrumental in drawing these two *sunni* empires closer to one another in matters of politics and religion. The political cooperation between the Timurid and Ottoman states also led to fruitful cultural exchange. Ottoman students went to study in Samarkand and Herat, and scholars and artists emigrated from Central Asia to the Ottoman State. It is very likely that the close associations with the Timurid lands cultivated by the Ottoman students who went to study there played an ongoing role in this exchange.

Those scholars who were patronized by the Timurid sultans were greatly disturbed by the conflict and political instability in the Timurid Empire following the deaths of Shahruh and his elder son Ulug Beg (d. 1449) — the latter ruler also a great scholar. These upheavals ultimately forced scholars to emigrate from Khorasan, Transoxania, and Azerbaijan to more peaceful lands, where scholarly and scientific activities were better appreciated. Only the Ottoman lands and India, which was then under the rule of another Turkish dynasty, the Mughals, had the conditions these scholars required to pursue their work. While some poets and physicians went to India, many prominent mathematicians and astronomers preferred to come to the Ottoman lands.

Ottoman institutions of learning and science

In the Ottoman Empire, the most important institution of religious, scholarly, and cultural activities was the *medrese*. When new lands came under Ottoman rule, the Ottomans immediately founded numerous *medreses* all over these lands, as well as sponsoring those that had already been established by previous rulers.

In the Ottoman period, the basic structure of *medrese* education remained in keeping with prior tradition, but the methods and the objectives of teaching and the scope of the curriculum with regard to the sciences underwent several important changes. The first Ottoman *medrese* was established in Iznik (Nicaea) in 1331 by Orhan Bey (1326-1362), the second Ottoman sultan. Thereafter, members of the Ottoman dynasty, statesmen, and prominent men of learning continued to found *medreses* throughout the vast expanse of the Empire. The continuity and development of these institutions was ensured by financially secure pious endowments (*waqf*s).

Table 1: Ottoman *medreses* in important centers and regions

	XIVth century	XVth century	XVIth century	Undefined period	Total
Iznik	4				4
Bursa	19	11	6		36
Edirne	1(Darüsshifa)	20	10		31
Istanbul		23	113	6	142
Anatolia	12	31	32	13	88
The Balkans	4	12	18	5	39
Syria			3		3
Hijaz			6		6
Yemen			1		1
Total	40	97	189	24	350

Table 2 : Distribution of *medreses* according to the reigning Sultans

Orhan Gazi (1326-1359)	10
I. Murad (1359-1389)	7
I. Bayezid (1389-1402)	23
Çelebi Mehmed (1402-1421)	7
II. Murad (1421-1451)	38
II. Mehmed (1451-1481)	30
II. Bayezid (1481-1512)	33
I. Selim (1512-1520)	8
I. Süleyman (1520-1566)	106
II. Selim (1566-1574)	17
III. Murad (1574-1595)	42
III. Mehmed (1595-1603)	5
Medreses in undefined periods	24
Total	350

Table 3 : Rumeli *medreses* in the Ottoman period

Regions	*Medreses*
Greece	189
Bulgaria	144
Albania	28
Bosnia-Herzegovina, Croatia, Montenegro	105
Kosova-Macedonia-Serbia, Slovenia, Voivodinia	134
Romania	9
Hungary	56
Total	665

In a short period of time, a large number of students were educated in the emerging Ottoman *medreses*, and an active scholarly environment was formed on the Ottoman soil, thanks to the instruction offered in the *medreses* by great men of learning. The tradition of sending the Ottoman scholars abroad and inviting scholars from other countries to Anatolia also continued. These travels for scholarly purposes are further evidence of the scientific dynamism of this period.

Active cultural and educational relations were established between Ottoman Anatolia and its Muslim neighbors : Iraq, Iran, and the Persian heartland of Transoxania in the east, and the Arab countries in the south, especially Syria and Egypt. Men of learning trained in these places were welcomed in the Ottoman world and their works were used as textbooks in the *medreses*.

Shortly after Byzantine Constantinople's final surrender to Sultan Mehmed II (" Fatih " or " The Conqueror ") in the year 1453, the Sultan built the first major religious and educational complex of Ottoman Istanbul. This was the Fatih *külliyesi* (Fatih complex), consisting of a mosque and courtyard surrounded by an elementary school (*mektep*), colleges of higher learning, a hospital, public kitchens, and auxillary buildings. The construction of this *külliye* set an example for the Sultan's successors and high-ranking members of the ruling class to follow. The colleges of the Fatih Complex (known as the *Sahn-i Semân medreses*) included 16 *medreses* (constructed adjacent to the mosque) : eight middle-level schools (*tetimme*) and eight higher level schools (*âlî*). This was an important breakthrough in the development of the *medrese* system in the Muslim world. Graduates of the Fatih *medreses* went on to serve as teachers, bureaucrats, *kadi*s (judges), *kazasker*s (highest ranking judges) and *mufti*s in the Ottoman Empire.

As a result of a general climate of political stability and economic prosperity, distinguished scholars and artists from all over the Islamic world assembled in the new Ottoman imperial capital of Istanbul. In addition to the favorable political conditions, the changes made in the traditional *medrese* system during the reign of Sultan Mehmed II fired the progress of Ottoman scientific education and production. With the foundation of the Fatih *medreses*, the Islamic world experienced an unprecedented wave of scientific progress. This was due in part to a broadening of the sope of *medrese* education, opening the door to the rational sciences, including subjects such as logic, mathematics, astronomy, and the natural sciences, in addition to continued instruction in religious sciences. For the first time — at least as regards the Ottoman *medrese* system — the charter of these colleges officially stipulated that the teachers should have some knowledge of *ulum-i akliye* (rational sciences) in addition to the traditional religious sciences. It is thus implied that the teaching of rational sciences was slowly being included within the formal *medrese* education. The increased interest in rational sciences also led to an increased production of scholarly works.

The institutions that were established under the patronage of Sultan Mehmed II and his son and successor Sultan Beyazid II were instrumental in the progress and development of Ottoman science in the second half of the fifteenth century. The continuing patronage given to scholars and the cultivation of a scholarly environment in the sixteenth century, during the reigns of sultans Selim I, Süleyman I, Selim II and Murad III, resulted in an active period in the history of Islamic science in Istanbul. Original contributions were made in, for example, Taqi al-Din's work in the field of astronomy. We shall return to this later.

When Egypt was conquered and annexed to the Ottoman Empire in 1517, many *sunni* lands in the Middle East and North Africa came with it into the fold of the Ottoman Empire. By the sixteenth century, when the Ottomans had become a world power, the Ottoman world was virtually synonymous with the Islamic world.

Establishment of the Süleymaniye Complex by Sultan Süleyman the Magnificent (r. 1520-1566) marked the final stage of development in the Ottoman *medrese* system. The charter of the Süleymaniye *medrese*s made specific reference to the teaching of rational sciences in the *medrese*s established by Sultan Mehmed II, and it, too, stipulated that these sciences should be taught during the course of formal education offered in these colleges. A specialized *medrese* named *Dâruttib* (medical college) was added to the conventional four *medrese*s. Thus, in addition to the *shifâhânes* (hospitals) where medical students were trained in the traditional hands-on way, an independent college was now established in the Süleymaniye Complex to give medical education. This college functioned until the middle of nineteenth century when European-style medical schools were opened. Other specialized *medrese*s established by the Ottomans were the *Darülhadîs* and the *Darülkurrâ*, the former being the highest institution in the hierarchy of the *medrese*s. As the pious foundations which provided the financial resources of the *medrese*s grew rich, scholarly and educational life developed to even greater heights.

In addition to formal scholarly education on offer in the *medreses*, there were other trade and professional institutions where sciences, including medical sciences and astronomy, were taught and practiced by the master-apprentice method. These scientific educational institutions were typically housed in the mansions of distinguished men of society, or in the *shifâhâne*s (hospitals), *muvakkithâne* (timekeeper's office), or the office of the *müneccimbashi* (chief astronomer).

The institutions in Ottoman society which provided health services and medical education were called *dârüsshifâ*, *shifâhâne*, *or bîmâristan*. The Seljuks of Anatolia had built *dârüsshifâ*s in Konya, Sivas, and Kayseri. Similarly, the Ottomans built several *dârüsshifâ*s in major cities such as Bursa, Edirne, and Istanbul. Some European sources report that there were a large number of *dârüsshifâ*s in Istanbul in the sixteenth and seventeenth centuries. These were not constructed independently but rather were part of a *külliye* (complex). The most important Ottoman *dârüsshifâ*s were the following : the Fatih *dârüsshifâ* in Istanbul (1470), the Beyazid II *dârüsshifâ* in Edirne (1481), the Süleymaniye *dârüsshifâ* in Istanbul (1550), the Haseki *dârüsshifâ* in Istanbul (1550), and the Hafsa Sultan *dârüsshifâ* (1522-23), built by the wife of Sultan Selim I in her name in the city of Manisa. Physicians were trained and patients treated in these *dârüsshifâ*s, some of which continued functioning right up to the middle of the nineteenth century, when European-style hospitals began to take over.

*Muvakkithane*s, or offices of the timekeepers, were public buildings located in the courtyards of mosques and *mescid*s in almost every town. They became widespread especially after the conquest of Istanbul by the Ottomans. They were run by the *muvakkit*, the person responsible for keeping the time. In particular, the *muvakkit* was responsible for keeping track of the correct prayer hours. Instruments used for timekeeping in the fifteenth and sixteenth centuries included the *rubu tahtasi* (quadrant), the *usturlâb* (astrolabe), the sextant, the octant, the hourglass, the sundial, the mechanical clock, and the chronometer. Depending on the timekeeper's expertise, the *muvakkithane*s might also function as the centers where mathematics and astronomy were taught.

In addition to the above-mentioned institutions, all of which were financed by *waqf* endowments, two institutions, the offices of the *hekimbashi* (Chief Physician) and the *müneccimbashi* (Chief Astronomer) were set up within the official Ottoman bureaucracy.

Office of *hekimbashi* (Chief Physician) was vested with full responsibility for the Sultan's medical needs, as well as those of the imperial family and the palace staff. However, he was more than just a private physician to the palace. In addition, the administration of all medical institutions and practitioners in the State, including physicians, surgeons, ophthalmologists, and pharmacists was placed under the direction of the office of the *hekimbashi*. The best-trained physicians were chosen from among the learned classes for this office. Some of the physicians who took the post of Chief Physician were among the most celebrated scholars of their time.

Also under the Office of the Chief Physician was the office of the *müneccimbashi* (Chief Astronomer), who was selected from among the religious intellectuals that graduated from the *medreses*. The institution of *müneccimbashi* was established in the late fifteenth or early sixteenth century. The office was entrusted with the preparation of official calendars, prayer and fasting timetables, and horoscopes for prominent statesmen and members of the imperial family. Until the year 1800, Ottoman calendars were made according to the *Zij* of Ulug Beg ; from that date onward Jacques Cassini's *ephemerides* were used.

The Chief Astronomer (or one of the other senior astronomers in the office) determined the auspicious dates and hours for important events, including imperial accessions, wars, imperial births, weddings, the launching of ships, and so on. The Chief Astronomer also kept track of unusual astronomical events (and the earthly cataclysms to which they were believed to be related), such as the passage of comets, earthquakes, fires, and solar and lunar eclipses. The Office of the Chief Astronomer forwarded information concerning such events to the palace together with their professional interpretations. The administration of the *muvakkithane* also came under the duties of the Chief Astronomer. The famous observatory founded in Istanbul in the late sixteenth century was administered by the Chief Astronomer Taqi al-Din Rasid (d. 1585). A total of thirty-seven individuals held the post of Chief Astronomer over the course of the Empire's history. The post of Chief Astronomer was retained all the way up until the Empire came to its final end ; it was abolished in 1924. In its place, the *bashmuvakkitlik* (office of the Chief Timekeeper) was established in 1927.

FOUNDATION OF THE ISTANBUL OBSERVATORY

Chief Astronomer Taqi al-Din was born to a family of Turkish descent in Damascus. He came to Istanbul from Egypt in 1570, and was appointed *müneccimbashi* by Sultan Selim II (1566-1574). Shortly after Sultan Murad III's (1574-1595) accession to the throne, he undertook the construction of an observatory in Istanbul. This was an elaborate structure which included, in addition to the observatory itself, residential quarters, offices for the astronomers, and a library. It was planned as one of the largest observatories in the Islamic world, and equipped with the best instruments of the time. Taqi al-Din's Istanbul observatory was comparable to Tycho Brahe's (1546-1601) Uranienborg Observatory, built in 1576. Indeed, there is a striking similarity between the instruments used by Tycho Brahe and those of Taqi al-Din.

Taqi al-Din had fifteen assistants in the Istanbul observatory. From his *zij* titled *Sidrat Muntaha'l-Afkâr*, we understand that he started making observations at the new observatory in the year 1573. It is generally agreed that the observatory was demolished on 4 Dhilhijja 987 in the Islamic calender, corresponding to January 22, 1580. Therefore, we may suppose that Taqi al-Din's observations were carried out between 1573 and 1580.

Taqi al-Din developed a new method of calculation to determine the latitudes and longitudes of the stars by using Venus and the two stars near the ecliptic, known as Aldebaran and Spica Virginis. By using the method called " three points observation " he calculated the eccentricity of the orbit of the Sun, and yearly mean motion of the apogee. According to Copernicus the eccentricity is 1p 56' ; according to Tycho Brahe it is 2p 9', and according to Taqi al-Din it is 2p 0' 34" 6''' 53'''' 41''''' 8''''''. As compared to modern calculation, Taqi al-Din's is the most accurate value. According to Copernicus, the annual motion of the apogee is 24" ; to Tycho Brahe it is 45", and to Taqi al-Din it is 63". Its real value is 61". As far as world astronomy is concerned,

Taqi al-Din's results can be said to be the most precise in the calculation of solar parameters.

As well as producing new observations and methods of calculation, Taqi al-Din invented a number of new astronomical instruments. The instruments in use in his observatory included the following : the armilary sphere, parallactic ruler, and the astrolabe, which were originally invented by Ptolemy ; the azimuthal and mural quadrants, which were invented by Muslim astronomers ; a *mushabbaha bi'l-manatic* (the sextant, an instrument with cords used to determine the equinoxes), built by Taqi al-Din, which very much resembled an instrument of the same type invented by Tycho Brahe. Taqi al-Din also built a wooden quadrant to measure azimuths and elevations, and mechanical clocks to measure the true ascension of the stars. The latter was one of the most important discoveries in the field of practical astronomy in the sixteenth century, because the clocks previously in use were never accurate enough. To quote Taqi al-Din, " We constructed a mechanical clock with three dials which show the hours, minutes, and seconds. We divided each minute into five seconds ".

Based on his own observations, Taqi al-Din prepared a *zij* for Sultan Murad III entitled *Sidrat Muntaha'l-Afkâr fi Malakut al-Falak al-Davvar - Zij-i Shahinshahi.* This *zij* contained the tables of the sun, but these were incomplete. His work titled *Jaridat al-Durar wa Kharidat al-Fikar* contained the tables of the moon, which were not based on his own observations. It must be noted that Taqi al-Din's observations were exact and he used the decimal fractions instead of the sexagesimal fractions in his astronomical calculations. He also prepared trigonometry tables according to decimal fractions.

Taqi al-Din, who read the works of Muslim scholars, produced a critique of early astronomical works in addition to reporting his own findings. Without doubt his works are representative of the most advanced stage of Ottoman science. Within a short time of its foundation, the Istanbul observatory was contributing the latest advances in the Islamic tradition of astronomy. Unfortunately, the observatory was demolished on account of false religious assertions made by envious statesmen — a dispute in which Taqi al-Din was caught between rival sides. The demolition of the observatory marks the beginning of the decline of a classical Ottoman scientific tradition.

Ottoman scientific literature in the classical period

The Ottoman scientific literature in the classical period was produced mainly within the milieu of the *medrese*. Original works and translations were produced in the fields of religious sciences, mathematics, astronomy and medicine, in addition to which a great number of textbooks were compiled by scholars. These works were written in Arabic, Turkish, and, to a lesser extent, Persian. (These were the three languages — known as the *elsine-i selâse* —

which the Ottoman scholars mastered). At first, Ottoman scientific literature was written mostly in Arabic, the basic and common language of Islamic civilization. However, in the fourteenth and fifteenth centuries, a movement was begun toward the production of Turkish translations of Arabic works. This activity was encouraged by those administrators who did not know Arabic, and in time Turkish was used to disseminate information to the public in their own vernacular language. Thus a concerted effort was made to translate works in every field of Islamic learning and practice (from encyclopedic manuals on medicine and drugs, geography, astronomy and the interpretation of dreams, to treatises on music and dictionaries) into simple and clear Turkish, and in this way to introduce Islamic culture to a wider public. From the eighteenth century onward, the majority of scientific works were written in Turkish, and upon the establishment of the first printing press in Istanbul in 1727, Ottoman Turkish became the most frequently used language in the transmission of knowledge about the modern sciences in the Ottoman Empire.

Bursali Kadizâde-i Rumî (d. 844/1440), the first prominent Ottoman scholar, made important contributions to the development of an Ottoman scientific tradition and literature. He started his scholarly career in Anatolia, where he wrote his first book. He then moved and settled in Samarkand. Among his works were *Sharh-Mulakhkhas fi'l Hay'a* (a commentary on the *Compendium on Astronomy* of Chaghminy) and *Sharh Ashkal al-Ta'sis* (a commentary on *The Fundamental Theorems of Geometry* of Samarkandî), both written in Arabic. He was made the chief instructor of the Samarkand *medrese* and the director of the observatory founded by Ulug Beg (d. 1449) in Samarkand. He was also the co-author of *Zij-i Gurgani (Zij-i Ulug Beg)*, the famous astronomical tables of Ulug Beg, written in Persian. He simplified the calculation of the sine of a one-degree arc in his work *Risala fi Istikhraj Jaybi Daraja Wahida* (*Treatise on the Calculation of the Sine of a One-Degree Arc*). Kadizâde-i Rumî's works made a significant impact on the Ottoman Empire, and his influence continued to expand through the works of Ali Kushcu (d. 1474) and Fathullah al-Shirvânî (d. 1486), two students of his who came to the Ottoman Empire from Turkestan.

In the preface of his work *Sharh Ashkal al-Ta'sis*, Kadizâde-i Rumî opined that the philosophers who pondered the creation and the secrets of the universe, the jurists (*faqih*s) who passed *fetvâ*s on religious matters, state officials, and *kadis* (judges) should all have knowledge of geometry. Thus, he emphasized the importance of studying mathematics for the pursuit of philosophical, religious, and worldly matters. This way of thinking reflects the general character of Ottoman science in the classical period — which lasted until the Ottomans began to adopt Western science. However, in the period of modernization, it became apparent that the Western concept of man's domination of nature through science and technology was foreign to Ottoman scholars, who came out of an Islamic scientific legacy and belief system.

Among other astronomy books read in this period were *Urjuza fi Manazil al-Kamar wa Tulu'iha* (*Poem on the Mansions of the Moon and their Rising*) and *Manzuma fi Silk al-Nujum* (*Poem on the Orbits of the Stars*), both written in Arabic by 'Abd al-Wahhab b. Jamal al-Din b. Yusuf al-Mardanî'. Two books by Nasir al-Din al-Tusî, the founder of the Maragha school, titled *Risala fi'l-Takwim* (*Treatise on the Calendar*) and *Si Fasl fi'l Takwim* (*Thirty Sections on the Calendar*) were translated from Persian into Turkish. Ahmed-i Dâ'î (d. *ca.* 825/1421) is known to be the translator of the second work.

In addition to Samarkand, Egypt was another important source for Ottoman science in the classical period. Haci Pasha (Celâleddin Hidir) (d. 1413 or 1417), a well-known physician educated in Egypt, wrote two books in Arabic, *Shifa' al-Askam wa Dawa'al-Alam* (*Treatment of Illnesses and the Remedy for Pains*) and *Kitab al-Ta'alim fi'l-Tibb* (*Book on the Teaching of Medicine*), both of which played an important role in the progress of Ottoman medicine. This author had many other works in both Turkish and Arabic.

In the field of medicine, the works of Sabuncuoglu Sherafeddin (d. *ca.* 1468) were particularly important in the development of Ottoman medical literature. His first book on surgery written in Turkish was titled *Jarrahiyat al-Khaniyya* (*Treatise on Surgery of the Sultans*) was an all-in-one handbook for every branch of medical science. The book was a translation of Abu'l Kasim Zahrawi's *al-Tasrif.* However, it was enriched by miniatures depicting surgical interventions. A chapter on the cauterization of boils and another on the treatment of herpes was added together with a list of drugs mentioned in the book. As the book includes the first miniatures depicting surgical operations, it is well-known in the history of Islamic medicine. In addition to classical Islamic medical information, the work reports on the author's own experiences and shows evidence of influence from Turco-Mongolian and far Eastern medical traditions. Sabuncuoglu's influence was felt outside the boundaries of the Ottoman Empire — particularly in Safavid Iran — through his student Ghyas Ibn Mohammed Isfahanî.

Sultan Mehmed the Conqueror, who is famous among the Ottoman sultans for his patronage of scholars, displayed an avid interest in classical Greek sources and the contributions of European culture to scholarship. He ordered the Greek scholar Georgios Amirutzes of Trabzon and his son to translate the *Geography* of Ptolemy into Arabic and to draw a world map based on it. Mehmed had first taken an interest in European culture when he was the crown prince living in the Manisa palace. In 1445, the Italian humanist Ciriaco d'Ancona and other Italian scholars resident in the Manisa palace taught Mehmed Roman and European history. Later, while Patriarch Gennadios prepared for Mehmed his work on the Christian faith, the *Itikadname* (*Book of Belief*), Francesco Berlinghieri and Roberto Valtorio presented him with their respective works *Geographia* and *De Re Militari.* These works are still extant in the collections of imperial manuscripts housed in the Topkapi Palace Museum.

Among the most important examples of Mehmed II's patronage of Muslim scholars and encouragement of scholarly writing, we may cite his request to the prominent Ottoman intellectuals Hodjazade and Ala al-Din al-Tusi that each produce a work comparing Al Ghazzali's *Tahafut al-Falasifa* (*The Incoherence of the Philosophers*), a critique of peripatetic philosophers regarding metaphysical matters, with *Tahafut al-Tahafut* (*The Incoherence of Incoherence*), which was Ibn Rushd's reply to this critique.

However, undoubtedly the most notable scientist of the Conqueror's time was Ali Kushcu (whose real name was Muhammed b. Ali), a representative of the Samarkand tradition. Ali Kushcu wrote twelve books on mathematics and astronomy, including a commentary on the *Zij-i Ulug Beg* written in Persian. His other two works in Persian are *Risala fi'l-Hay'a* (*Treatise on Astronomy*) and *Risala fi'l-Hisab* (*Treatise on Arithmetic*). He rewrote these two works in Arabic with some additions — and under new titles, as *al-Fathiyya* (*Commemoration of the Conquest*) and *al-Muhammadiyya* (*Dedication to Sultan Mehmed*). Both of these works were taught in the Ottoman *medreses*.

A noteworthy scholar of the Bayezid II period was Molla Lutfî (1481-1512). Molla Lutfî wrote a treatise in Arabic about the classification of the sciences, entitled *Mawdu'at al-'Ulum* (*Subjects of the Sciences*), and compiled a book on geometry titled *Tad'if al-Madhbah* (*Duplication of the Cube*). Part of the latter was translated from Greek.

Mîrîm Çelebi (d. 1525), the grandson of Ali Kushçu and Kadizâde-i Rumî, was another famous astronomer and mathematician of the early sixteenth century. He contributed to the development of Ottoman scientific tradition particularly in the fields of mathematics and astronomy, and was famous for his commentary on the *Zij* of Ulug Beg.

Interesting contributions to Ottoman scientific literature were made by emigrant Muslim and Jewish scholars from Andalusia. Ilya b. Abram al-Yahudi is one such scholar. Al-Yahudi settled in Istanbul during the reign of Sultan Bayazid II. There, after converting to Islam, he changed his name to 'Abd al-Salam al-Muhtadi al-Muhammedi and authored several medical and astronomical works in Arabic. In a book dedicated to Sultan Bayezid II, which he first wrote in Hebrew then translated into Arabic in 1503, al-Yahudi introduced an astronomical instrument of his own invention called *al-dabid*, claiming that it was superior to the *dhat al-halak* (armillary sphere) invented by Ptolemy. This treatise illuminates an aspect of Ottoman scientific literature which is not yet well known.

The production of scientific literature advanced considerably during the reign of Sultan Süleyman the Magnificent. Two major books written in Turkish on mathematics are known from this period, the *Jamal al Kuttab wa Kamal al-*

Hussab (*Beauty of the Scribes and Perfection of the Accountants*) and *'Umdat al-Hisab* (*Treatise on Arithmetic*), both by Nasuh al-Silahî al-Matrakî (d. 971/1564).

With their continuously expanding territorial empire and concurrent quest to dominate navally in the Mediterranean, Black Sea, and Red Sea, and Indian Ocean, the Ottomans had a clear interest in excelling in geography. Both the geographical works of classical Islam and the more contemporary European works were very useful in this respect. The first source of Ottoman knowledge about geography came from the Samarkand tradition. Meanwhile, by supplementing existing geographical knowledge with their own fresh observations, the Ottoman geographers produced original works.

During the sixteenth century, noteworthy contributions to Ottoman geography were made by the prominent naval captain Pîrî Reis, whose earliest surviving map was drawn in 1513. It is a *portolano*, without lines of latitude or longitude, but with careful delineations of coasts and islands. This was once a large world map ; however, only a part of it has survived. This fragment depicts southwestern Europe, northwestern Africa, and the eastern coasts of South and Central America. It is considered most astonishing for its early delineation of the eastern coastline of the Americas. The map was based on Pîrî Reis' experience as a sailor, as well as on other Islamic and European source maps — including, Pîrî Reis claimed, some of Columbus' own early maps of the Americas.

Pîrî Reis presented yet another world map to Sultan Süleyman the Magnificent in 1528. Of this map, only the part which contains the North Atlantic Ocean and the newly discovered areas of North and Central Americas has survived. Pîrî Reis also wrote the famous *Kitâb-i Bahriye* (*Book of the Sea*) which he presented to Sultan Süleyman the Magnificent in 1525. This important work consists of drawings and maps of the Mediterranean and Aegean coasts, and gives extensive information about navigation and nautical astronomy. In writing this book, he made use of both classical Islamic and contemporary geographical sources, as well as his own observations.

Another Ottoman naval officer, Admiral Seydî Ali Reis (d. 1562), a prominent figure in the field of maritime geography, wrote a noteworthy book in Turkish titled *al-Muhit* (*The Ocean*). His work contains his observations concerning the Indian Ocean, and astronomical and geographical information necessary for long sea voyages.

Nasuh al-Matrakî's *Beyan-i Menâzil-i Sefer-i Irakeyn* (*Description of the Places on the Way to Iraq During the Two Campaigns*) is a good example of descriptive geography.

Tarih-i Hind-i Garbî (*History of Western India*) is yet another geographical work produced in the sixteenth century. It contains information about geographical discoveries in the New World. This anonymous work based on Spanish and Italian sources was presented to Sultan Murad III in 1583. The work is in three parts. The third part relates to the travels and adventures of Columbus, Balboa, Magellan, Cortés, and Pizarro in the first sixty year after the discovery of America, from 1492 until 1552. Works such as this (and Pîrî Reis') demonstrate that the Ottomans were aware of the geographical discoveries made by the Europeans.

Geographical and historical ties with Europe were crucial in making the Ottoman Empire the first non-Western environment where Western science and technology spread. Geographical proximity allowed the Ottomans to learn about European innovations and discoveries. From the fifteenth century on, the Ottomans began to adopt European technologies, expecially those that concerned firearms, cartography, and mining. They also had access to Renaissance advances in astronomy and medicine through their contacts with Jewish scholars who emigrated to the Ottoman lands. However, Ottoman interest in European advances was selective, partly because of their feelings of moral and cultural superiority over Europe, and partly because of the self-sufficiency of their economic and educational system. They therefore did not keep close track of the scientific and intellectual developments of the Renaissance and the Scientific Revolution as they were unfolding. It is anachronistic to claim, as some modern historians do, that the Ottomans failed to realize that these developments would challenge them in the future.

Starting in the sixteenth century with the arrival of some European physicians and the spread of various diseases from the West, new medical ideas and methods of treatment and prophylaxis were introduced into the Ottoman Empire. In the seventeenth century, the medical doctrines of Paracelsus and his followers were seen in Ottoman medical literature under the names of *tibb-i cedid* (new medicine) and *tibb-i kimyaî* (chemical medicine). One of the most famous followers of this trend was Salih b. Nasrullah (d. 1669). In his work titled *Nuzhatu'l-abdan*, Salih mentioned the new medicines described in various European sources, and disclosed the composition of their remedies. Al-Izniki (18th century) likewise cited Arab, Persian, Greek, and European physicians in his *Kitab-i Kunuz-i Hayat al-Insan Kavanin-i Ettibba-yi Feylesofan*, and presented information on the new medicines as well as the old known ones. Ömer Shifaî (d. 1742), too, in his work *al-Cevher al-Ferid,* claimed that he had translated the book from a European language into Turkish containing remedies taken from the works of Latin physicians. Thus, new information on the composition and use of European medicines was disseminated and put into practice alongside traditional medicines until the beginning of the nineteenth century.

The famous Ottoman scholar and bibliographer Kâtip Çelebi (d. 1658), also known by the name of Haci Halife, was one of the first intellectuals to notice the widening gap between scientific advancement in Europe and the Ottoman world. He was able to approach both classical Islamic culture and contemporary European culture in an analytical frame of mind. He wrote books in Arabic and Turkish on a variety of subjects. With the help of a European convert, he translated the *Chronic* of Johanna Carion from Latin and titled it *Tarih-i Firengî Tercümesi* (*Translation of European History*). This book was first published in German in 1632, with numerous later editions and translations. He also compiled *Tarih-i Konstantiniye ve Kayasira* (*History of Constantinople and the Emperor*) which is also called *Ravnak al-Saltana* (*Splendor of the Sultanate*). Kâtip Çelebi's source for this work was evidently one of the reprints of the *Corpus universae historiae praesertim Byzantinae* (Paris, 1567), in this case one titled *Historia rerum in Oriente gestarum ab exordio mundi et orbe condito ad nostra haec usque tempora* (Frankfurt, 1587). The bulk of this book consisted of the histories written by Johannes Zonaras, Nicetas Acominatus, Nicephorus Gregoras, and Laonicus Chalcocondyles.

In the field of geography, Kâtip Çelebi translated the *Atlas Minor* of Mercator and A.S. Hondius under the title *Lawami' al-Nur fi Zulmat Atlas Minur* (*Flashes of Light in the Darkness of Atlas Minor*). Later, in his work *Mizan al-Hakk fi Ikhtiyar al-Ahakk* (*The Balance of Truth and the Choice of the Truest*), Kâtip Çelebi critiqued the intellectual life of his day.

Starting in the seventeenth century, there was a massive influx of translations of Western scientific works to the Ottoman Empire. Now we shall attempt to follow the introduction of modern scientific concepts as they were absorbed into the Ottoman scientific milieu. As far as we could establish, the first work of astronomy translated to Ottoman Turkish from a European language was the *Ephemerides Celestium Richelianae ex Lansbergii Tabulis* (Paris, 1641), the astronomical tables by the French astronomer Noel Duret (d. *ca.* 1650). Ottoman astronomer Tezkereci Köse Ibrahim Efendi (of Zigetvar) translated this book in 1660 under the title *Secencel el-Eflak fi Gayet el-Idrak*. This translation was also the first book in Ottoman literature to mention Copernicus and his heliocentric system. The initial reaction of the Chief Astronomer to this notion was to declare the book a " European vanity " ; but after learning its use and checking it against Ulug Beg's *Zij* (astronomical tables), he realised its value and rewarded the translator. His first reaction, however, typifies Ottoman reluctance to acknowledge European scientific superiority.

Ottoman astronomers of the classical period viewed heliocentricity, which aroused such controversy in Europe, as an alternative technical detail and did not make it a subject of any great polemics. One likely explanation for this nonchalance is that the heliocentric theory of the universe did not conflict with any religious dogmas. Subsequently, most astronomy books translated from European languages dealt with astronomical tables and did not touch upon the subject of heliocentricity again.

Among the translations completed in the second half of the seventeenth and early eighteenth centuries was a major book on modern geography by Abu

Bakr b. Behram el-Dimashki (d. 1691), based on Janszoon Blaeu's *Atlas Major* (1685).

Ibrahim Müteferrika translated Andreas Cellarius' *Atlas coelestis* (1708) from Latin, completing the translation in 1733 and calling it *Mecmuatü'l-hey'eti'il-kadime ve'l-cedide*, or " collection of old and new astronomy ". In 1751, Osman b. Abdulmennan translated Bernhard Varenius' work *Geographia Generalis* from Latin, calling it *Tercüme-i Kitab-i Cografya* (*Translation of Geographia Generalis*). Alongside these new translations, classical Ottoman astronomy, geography, and related scientific activities continued within their traditional framework.

A survey of Ottoman astronomical literature thus shows that after overcoming their feelings of superiority, Ottoman scholars readily accepted new information, techniques and concepts from Europe. The Ottoman administration looked favorably on such influences, and the religious scholars (*ulema*) were not particularly hostile — as evidenced in their attitude toward the theory of heliocentricity. There was no obvious conflict between religion and Western science at this stage.

Towards modernisation

As the seventeenth century drew to a close, the Ottomans, faced with European military and economic superiority, recognized the need to reform classical institutions. Yet they seemed unable to adapt to the changing conditions. They failed to understand modern economic problems, and remained bound to the traditional formulae defining the Near-Eastern political state. The weakness of the Ottoman army became apparent when it was defeated at Vienna in July 1683. Following this defeat, the Ottomans grudgingly began to acknowlege the superiority of the West in certain fields. The Treaty of Carlowitz (1699) impeded further Ottoman territorial advances in Europe, and it became painfully clear that Ottoman power was waning and Europe gaining the upper hand. Thus the Ottoman Empire embarked on a new era, one where leaders would look more attentively to the West as they tried to address some of Empire's problems.

Since the foundation of the Empire, the Ottomans had been keenly interested in European military and technological novelties, which they had selectively adopted for themselves. However, at the beginning of the eighteenth century, Ottoman-European relations entered a new, more intense, period. Confronted with the swift advances in European technology and the defeat of the Ottoman armies by the Europeans, the Ottomans became convinced that in order to master the techniques of modern warfare, they needed to open new channels of communication and technology transfer.

Then, during the first quarter of the eighteenth century, the Ottomans won decisive victories over the Persian and Russian armies in the east, as well as the Venetian armies in the west. Following these victories, a relatively long period of peace known as the *Lâle Devri* (Tulip Period) began. This period is most closely associated with the reign of Sultan Ahmed III (r. 1703-1730) and his Grand Vizier Nevshehirli Damat Ibrahim Pasha. It lasted until the Patrona Halil rebellion in 1730. During this period, the Ottomans were intensely interested in European ways of life and innovations. A new era began in Ottoman cultural life : novelties such as the printing house and pumping equipment for the firemen were imported from Europe. Meanwhile, close contacts between the high officials of the Ottoman state and European ambassadors were established.

Ottoman administrators followed new military developments closely after the year 1730. They trained officers and privates in modern techniques of warfare and equipped the army with the same weapons used by the European armies. In this way, they tried to restore the military balance with Europe. With this objective in mind, they established institutions of military education where new information and technological developments were taught. These schools stood in contrast to the traditional style of military training based on the master-apprenticeship relationship — an important new development in the Ottoman educational system as well as the military.

The first efforts to give modern technical training in the Ottoman military began in the *Ulufeli Humbaraci Ocagi* (The Corps of Bombardiers). The Corps was organized in 1735, with the efforts of Claude Alexander Comte de Bonneval (1674-1747), a general of French origin who had requested asylum from the Ottomans in 1729 and converted to Islam, taking the name of Ahmed. He was later known as Humbaraci Ahmed Pasha. The fact that he took refuge in Istanbul was a blessing for the Ottomans, who wished to make military reforms and could benefit from his expertise.

The Corps was divided into three subdivisions or *oda,* each consisting of twenty-five officers of different ranks. Some of these had administrative duties, while others handled military, educational, and medical affairs. The Corps was under the direct supervision of the commander-in-chief of the military and the Grand Vizier. The salaries of the officers and soldiers of the Corps of Bombardiers were paid from a special fund. A candidacy system was introduced and the salaries were fixed according to rank. The retirement pension of the bombardiers was also guaranteed.

The teaching staff of the Corps consisted of Ottoman as well as French and Scottish teachers. Mehmed Said Efendi was the instructor in geometry, Istanbullu Ibrahim Hodja taught arithmetic, geometry, drawing and the measurement of altitude, and Cenk Mimarbashi Selim Aga taught fortification, artillery, and mechanics. Classical Ottoman books on mathematics and European books were used concurrently. Thus, for the first time, mathematical sciences were included in the curriculum of a military institution and a new system of military training was introduced to Ottoman education. Before the foundation of the Corps, mathematical sciences were taught either in the *medrese* or in private tutoring. The military school thus opened a third venue.

After Ahmed Bonneval Pasha's death in 1747, the Corps of Bombardiers was administered for a short time by his adopted son Süleyman Aga, but it was soon dissolved. Nonetheless, Ottoman administrators redoubled their efforts to train army officers in the new military techniques, and in the 1770s new attempts were made to establish more enduring military-technical schools.

THE IMPERIAL SCHOOL OF MILITARY ENGINEERING

The foundation of the *Hendesehane* (literally " house of geometry ") within the Imperial Maritime Arsenal on April 29, 1775, represented a significant step forward in Ottoman military education. Referred to as the *Ecole de Théorie et de Mathématiques* in French documents, the *Hendesehane* was founded under the supervision of Baron de Tott, a French officer of Hungarian origin who came to Istanbul in 1770 and later entered the Sultan's service as a military adviser. The goal of the *Hendesehane* was to provide the imperial naval fleet (*donanma-i humayun*) with officers trained in the sciences, particularly geometry and geography. The French technician Sr. Kermovan and the Scotsman Campbell Mustafa taught mathematics at the *Hendesehane* until September, 1775, then Sr. Kermovan left Istanbul and Baron de Tott lost interest in this subject. With a regulation dated 1776, the *Hendesehane* became the first Ottoman institution where mathematics and the art of fortification were taught based on European sources, theories, and methods. After Baron de Tott left Istanbul in 1776, Cezayirli Seyyid Hasan, Second Captain in the Ottoman Imperial Fleet, was appointed to the post of *hodja* (professor) in the *Hendesehane*. Soon afterward, this institution was reorganised to conform to the traditional Ottoman bureaucratic structure.

After 1781, the *Hendesehane* was also called the *Mühendishane* (literally meaning " house of geometricians or engineers "), and ten students were trained there at a time. During the Grand Vizier Halil Hamid Pasha's term of office (1784), two military engineers, Lafitte-Clavé and Monnier, who were sent by the French Government to assist in reforming the Ottoman army and strengthening the fortifications, also taught in the *Mühendishane*, training officers in artillery, navigation and fortification.

As in the case of the Corps of Bombardiers (1735), Ottoman teachers were graduates of the *medreses* (and thus of the *ulema* class). Until the end of the eighteenth century both classical Ottoman books on science and foreign (mainly French) texts were used, particularly in the teaching of mathematics, astronomy, firearms, techniques of fortification, warfare, and navigation. After 1788, when the French left the Ottoman Empire, the *medrese* teachers took over responsibility for all courses at the *Mühendishane*.

As part of a series of new military reforms undertaken by Sultan Selim III (r. 1789-1807), a new *Mühendishane (Mühendishane-i Cedide)* was established in 1793 within the barracks of the Corps of Bombardiers. Here cannoniers, bombardiers, and sappers were to be trained. Courses commenced at the new *Mühendishane* in 1794 with a new generation of Ottoman engineer-teachers, among whom we may note the First Chief Instructor Hüseyin Rifki Tamanî, who had learned the art of fortification from French engineers like Lafitte-Clavé and Monnier in the old *Mühendishane*. The organization of this new institution was modeled on the other *Mühendishane*, and the teaching staff consisted of a professor (*hodja*), four assistant professors (*halifes*), and other functionaries. Here, members of the Corps of Bombardiers and Sappers were taught geometry, trigonometry, measurement of elevation, and surveying.

Between the years 1801 and 1802, a number of students selected from the Corps of Bombardiers, Sappers, and Architects were admitted to the *Mühendishane-i Cedide*. The teaching staff was composed of a professor and five assistant professors, and the students numbered about one hundred. In 1806, upon the issuing of an imperial decree of Sultan Selim III, a new regulation was prepared for this institution. Thereafter, it was called *Mühendishane-i Berri-i Hümayun* (The Imperial School of Military Engineering), a name which it retained for quite a long time.

The new regulation of 1806 reflected both European and Ottoman influences. The European practice of taking successive classes (four classes with four teachers, one of them being the chief instructor) was introduced for the first time in Ottoman educational system, while the practice of moving up from one class to a higher class (which was only possible if there was a vacancy in the senior class) continued in accordance with the old Ottoman administrative tradition. This Ottoman system of advancement was called *silsile* (chain).

In 1793, the first *Mühendishane*, located in the Imperial Arsenal, was transformed into a school where shipbuilding, navigation, surveying, and geography would be taught. The French naval engineer J. Balthasar Le Brun was put in charge of the school's administration. After he returned to France, he was succeeded by the Ottoman naval officers whom he had trained. Again by imperial decree of Sultan Selim III, this school officially became known as the *Mühendishane-i Bahri-i Humayun* (The Imperial School of Naval Engineering) and served the Empire for many years under this name.

New Military Schools of the Nineteenth Century

Institutions of Modern Medical Education

Until the beginning of the nineteenth century, the *Süleymaniye Tip Medresesi* (Süleymaniye *Medrese* of Medicine), which was founded by Sultan Süleyman the Magnificent as part of the Süleymaniye Complex in the sixteenth century, had been the only *medrese* devoted to medical education. However, this should not be taken to mean that it was the only Ottoman institution for training physicians. Many physicians were trained in the hospital, or *darüsshifa*, themselves. In addition to these, there were some non-Muslim Ottoman subjects who had studied medicine in Europe, and Jewish physicians from Spain who had taken refuge in the Ottoman Empire.

Many state institutions like *Tophane-i Amire* (Imperial Gun Foundry), *Tersane-i Amire* (Imperial Maritime Arsenal) and other military organizations had their private physicians and surgeons. Although we know little about Ottoman medical training in the classical period, we believe that it was in most cases based on a master-apprentice tradition. In the eighteenth and nineteenth centuries, as education and professional life were modernized, medicine was among the fields that were given top priority. As in many other fields, the first attempts to introduce modern medical practices were made in the military. In order to train physicians and surgeons for the Imperial Maritime Arsenal (*Tersane*), a school called the " *Tersane* School of Medicine " was opened in January, 1806. The aim of this school was to teach modern medicine, and in so doing to increase the number of Muslim physicians in the Empire. The courses were to be given in French or Italian — the languages most commonly used by the Ottoman Levantine physicians and pharmacists who were in contact with Europe — and would use European textbooks.

Although this school was closed only two years later, in 1808, the significance of its establishment within the Imperial Maritime Arsenal must be noted. As we have already seen above, the Arsenal was an important conduit through which modern science and technology were intoduced into the Ottoman Empire. Three major streams of early modernizing influence may be identified in connection with the Imperial Maritime Arsenal : the first was shipbuilding — a shipbuilding based on modern methods and using new technologies, which were first introduced in the Ottoman Empire at the Arsenal. Second was the teaching of mathematics, astronomy, and engineering at the *Hendesehane* and *Mühendishane*, the two schools located within the Arsenal. Third was the introduction of modern medicine and medical education at the Arsenal medical school.

Toward the beginning of the nineteenth century, two personalities were particularly influential in shaping Ottoman medical education. The first was Shanizade Mehmed Ataullah Efendi (d. 1826), a scholar of many interests and a valuable encylopedist, trained in the sciences as well as in European languages. In his well-known five-volume medical work entitled *Hamse-i Shanizade*, modern medicine and anatomy were presented to Ottoman readers in a comprehensive form for the first time.

The second scholar was Mustafa Behçet Efendi (d. 1834), considered the founder of modern medical education in Turkey. The *Tiphane-i Amire* (Imperial School of Medicine) was founded in 1827 while Behçet Efendi held the post of Imperial Chief Physician. Like the earlier Arsenal medical school, the goal of this new medical school was primarily to train physicians and surgeons for the army. In 1831, a separate school for training surgeons, the *Cerrahhane-i Amire* (Imperial School of Surgery) was opened at the Gülhane gardens, adjacent to the Sultan's Topkapi Palace. In 1836, the medical school was transferred next door to the school of surgery, and the curriculum was redesigned under the direction of the French surgeon Sat de Gallière. In 1838, these two schools (medicine and surgery) were combined in a single institution named *Mekteb-i Tibbiye-i Adliye-i Shahane* (The Imperial School for Medical Sciences) and moved to the Galatasaray district of Istanbul, where they would be housed in a single building. C. Ambroise Bernard, a young physician from Austria, was appointed director to the school in the same year. Under Dr. Bernard, medical education in the Ottoman Empire entered a period of stability, both in terms of content and method. French became the language of instruction. The practice of promoting students depending on the vacancies in the senior classes (the above-mentioned " chain " system) was abolished and program of study limited to five years, at the end of which graduates received diplomas certifying completion of the course of study. This European style of academic promotion was also adopted in the Imperial School of Military Engineering, where institutional certification of learning and graduation replaced the more personalistic process of classical system, known as *icazet.* In that system, a student was given a license, or *icazet*, by his master when the latter deemed him prepared. The master was conceived of as a personal authority in a chain of transmission of (religious) learning and scholarly authority reaching back to the Prophet, the first source of divine knowledge.

Following the proclamation of the Imperial *Tanzimat* Rescript in 1839, non-Muslim subjects were allowed to enroll in the Imperial School of Medicine for the first time. With this change the number of Muslim students gradually decreased. Non-Muslim students proved more successful in their studies — and particularly in mastering French, because of their affinity with European culture. As a way of redressing this imbalance, Cemaleddin Efendi, the Superintendent of the Imperial School of Medical Sciences, set up a special class called the " distinguished class " (*mümtaz sinif*) where special emphasis was given to the teaching of Turkish, Arabic, and Persian languages. This class became the nucleus of the civilian school of medicine that was founded in 1867, and paved the way for the spread of modern civilian medical education among the Ottomans. Under the leadership of Kirimli Aziz Bey and his colleagues, who were all graduates of this special class, a *Mekteb-i Tibbiye-i Mülkiye* (Civilian School of Medical Sciences) was opened as a part of the Imperial School of Medical Sciences. Here Turkish was the language of instruction. Thus, civilian medical education became independent from military medical education. However, shorty after, in 1870 a decision taken by the *Darüshshura-i Askerî* (Medical Council of the Ministry of Defence) decreed that instruction in the Imperial School for Medical Sciences, as well, would thenceforth be given in Turkish. This decision provoked heated debates between some Turkish physicians and their non-Muslim Francophone colleagues. The Turkish physicians prevailed, and from that point on medicine in the Ottoman Empire was taught exclusively in Turkish. It should be noted that this decision applied to Ottoman provincial medical schools as well as those in Istanbul. Thus, for example, even in the Ottoman Medical School in Damascus, where the local language was Arabic, the language of instruction was Turkish.

As the number of students in the civilian medical school increased, the school moved to a larger campus in 1873. In 1909, both the military and civilian medical schools were moved again, this time to a new and larger building in the Haydarpasha district of Istanbul. Then, in 1915, the civilian school of medicine was annexed to the *Istanbul Dârülfünun* (university) as the Faculty of Medicine. This faculty at the *Dârülfünun* became the first among all subsequently-established Turkish faculties of medicine. This building is still in use, and since 1980 houses the Marmara University Faculty of Medicine.

The Imperial Military School

In 1826, Sultan Mahmud II (r. 1808-1839) abolished the Janissary Corps, which had until then been one of the cornerstones of the Ottoman army. In its place he formed a new corps called *Asakir-i Mansure-i Muhammediye* (Mohammed's victorious soldiers). In 1831, plans were drawn up for a military school that would train officers for this new force and educate them in modern techniques of warfare. Until then, officers had been trained in the Imperial School of Military Engineering.

The Ottomans founded this school on the model of the French *Ecole Militaire*. The school, which had a capacity of 400 students, opened its doors in 1834 under the name of *Mekteb-i Harbiye-i Shahane* (Imperial Military School) and was headed by Namik Pasha, a general who was educated in Europe and well versed in European languages. The organization of the Imperial Military School differed from the schools of engineering and medicine in several respects : it consisted of eight classrooms where different courses could be held separately. Teachers from both the schools of military and naval engineering collaborated in preparing a curriculum that would improve the general level of education. As part of this program, students and officers would be sent to Vienna and Paris for training — in part so that the demand for teachers in the new Imperial Military School might be met.

In 1838, Emin Pasha was appointed as the superintendent of the Imperial Military School. He divided the school into two sections. The first was a senior level called *Mekteb-i Fünun-i Harbiye* (School of Military Arts) which had a four-year program of training, and the second was junior level program of three years called *Mekteb-i Fünun-i Idadiye* (Preparatory School for Military Arts). During Emin Pasha's tenure, the number of the teaching staff was increased and instruction was modernized by a combination of European teachers and young Ottoman officers educated in Europe. In 1848, Kimyager Dervish Pasha, who succeded Emin Pasha as superintendent of the school, prepared a new set of regulations inspired by those at the French *Ecole de Saint Cyr.* These regulations re-organized the Imperial School of Military Arts, making it conform completely to the European model. A school of veterinary sciences *(Baytar Mektebi)* to meet the army's need for veterinarians was established in 1846 as part of the Imperial School of Military Arts.

The Emergence of Secular Civilian Institutions in the Ottoman Educational Life

At the outset of modernization in Ottoman scientific and educational life, during the pre-*Tanzimat* period (the end of the eighteenth and beginning of the nineteenth centuries), there were two kinds of educational institution in the Ottoman Empire : the *medrese* and the institutions of military and technical education (engineering, medical, and military schools).

Institutions of military education were established and developed within the framework of military reforms that began in the eighteenth century, as described above. The modern education provided by these institutions was different from the classical Ottoman style of the *medreses*, and led to the emergence of a new understanding of education. As previously mentioned, this transformation was first encountered in the schools of engineering, where there was a shift from the classical Islamic tradition to the Western tradition. For a while, the two approaches coexisted, but from the second half of the nineteenth century onward, a conflict began to emerge between " the old " and " the new " — often conceived of as a conflict between " religion " and "science ".

The schools of engineering are often seen as the first centers of Western-style education. Yet in some ways, they were actually examples of an Ottoman-Western synthesis. The main purpose of the institutions of military and technical education was described as follows : *" Mühendishanelere fünun-i berriye ve bahriyeden hendese, hesap ve cografya fenlerinin intishari ve Devlet-i Aliyye'ye ehemm ve elzem olan sanayi-i harbiyenin talim ve teallümü ve kuvveden fiile ihraci "*. Briefly, in the language of the time, the objective was to train the *mütefennin zabit* (military officers educated in the military sciences). The statesmen and scholars of the period believed that a regular army equipped with modern technology could save the country. This was the reason why they turned to the West : in order to train and equip the officers with modern sciences and technology.

The Imperial *Tanzimat* Rescript (1839) did not explicitly describe any specific goals with regard to education and learning. However, before long it was clear to all that reforms in the educational system as a whole were a necessary step in modernization. In January of 1845, when Sultan Abdülmecid visited the *Meclis-i Vâlâ* (Supreme Council of Juridical Ordinances), he pointed to the dire necessity of fighting ignorance in all aspects of life. In particular, he saw education as a means to improve the public works of the Empire and thereby prosperity of its people. He ordered that priority should be given to the education of the people.

Thus for the first time in Ottoman history, the State started considering the subject of " educating the public " — including the establishment of a Central Office to coordinate such a project. In 1845, a Temporary Council of Education planned a system of education based on European (and particularly French) models. The model called for three stages of education, *primary*, *intermediate*, and *high*. The *intermediate* was accepted by the Council as a school where young school boys would be trained for the *darülfünûn* (university), thus securing their place in the Ottoman educational system. As this intermediate training proved effective, the number of intermediate schools increased from four to ten between the years 1847 and 1852.

At this time, Ottoman administrators found that they had to recruit newly-trained bureaucrats and officials for the new departments created by reform in the Central Administration. Throughout the nineteenth century, trained civilians were needed in the fields of agriculture, animal breeding, forestry, mining, engineering, industry, law, and the fine arts. As a first step, the State devised a system that would help fill these posts. The result was modern schools, modeled on those in Europe, whose graduates would then be employed in the government services. The second step was finding the teaching staff that would train this new generation. A program was developed to solve this problem. This program dealt with middle-level schools such as the *Rüshdiye, Idadi,* and *Sultani,* establishments where male teachers who had been trained in the *Darulmuallimin* (teacher training school for men, established in 1846) and the *Darulmuallimat* (teacher training school for women, established in 1869) would teach. European teachers or Ottomans who had been trained in Europe were employed for teaching of the subjects listed above.

Sending Students to Europe for Training

The practice of sending Ottoman students to Europe for training began during the reign of Sultan Mahmud II. In 1830, four Ottoman students named Hüseyin, Ahmed, Abdullatif, and Edhem, who were studying in the *Enderun* (Palace School), were sent to France to study military subjects under the patronage of Serasker Hüsrev Pasha. Their expenses were met by the Imperial Treasury (*hazine*). In 1839, thirty-six students were selected from the military and engineering schools and sent to London, Paris, and Vienna to learn European technology. On their return, they would be employed in the Ottoman military factories and workshops where heavy industry and technology were used

— i.e. the arsenal, gunpowder works, rifle and cartridge factories, and the foundry.

Following the proclamation of the *Tanzimat*, eligible students who wished to receive training in the civilian services were also sent to Europe with funding from the state. Because of the influx of non-Muslims into the new educational institutions at this time (as discussed above), in the year 1840 a significant number of non-Muslims were among the students sent to Europe. The ratio of the Muslim to non-Muslim students varied in different periods. Between 1848 and 1856, about fifty students went to Paris, this number rose to sixty-one between 1856 and 1864. An Ottoman school by the name of *Mekteb-i Osmanî* was opened in Paris in 1857 to educate the Ottoman students in the arts and sciences and thus enable them to follow the classes offered in French schools. Until 1864, all Ottoman students in Paris attended this school. Between 1864 and 1876, a total of ninety-three students were sent to Paris ; forty-two of these went to study science, and fifty-one to receive apprenticeship training in various technical vocations. On January 13, 1870, twenty students from the *Mekteb-i Sanayi* (The School of Arts and Crafts) went to Paris to study in various branches of arts. Thus, the number of students that were sent to Europe gradually increased over the course of the nineteenth century. In addition to military education, training was given in the professions and in arts needed for civilian purposes.

Darülfünûn (University)

In the history of Ottoman education, the *Darülfünûn* (which can best be translated into European terminology as " university ") stands as a unique institution in civilian higher education. It had no counterpart in the classical Ottoman system, since it was quite different from the *medrese*. The idea of a *darülfünûn* emerged during the *Tanzimat* period in response to the need to educate a greater number of students in a non-military form of higher education. Towards the middle of the nineteenth century the first attempt was made to found a *darülfünûn*.

The Council of Public Instruction stated that the aim of the *Darülfünûn* was to train future employees and cultivate enlightened civil servants (*münevver bendegân*), who would go on to serve the State in the best possible way. This training included the study of modern sciences. In November 1846, a contract was signed with G. Fossati, a Swiss architect of Italian descent, to build the imposing three-story *Darülfünûn* building in the Ayasofya quarter of Istanbul. The building would have 125 rooms and would resemble large European universities. Unfortunately, construction dragged on for many years. Yet in 1863, by the order of Grand Vizier Keçecizade Fuad Pasha, a program of public lectures was started in the incomplete building. When the building was finally completed in 1865 it was deemed too big for the intended purpose ! It was promptly turned over to the Ministry of Finance, and a smaller building was planned for the *Darülfünûn*. In order not to interrupt the ongoing lectures, the government rented temporary quarters in the mansion of Nuri Pasha in the Çemberlitas district. (This building was sadly destroyed in the Hocapasa fire). In 1869, the new, smaller building for the *Darülfünûn* was completed and classes began.

That same year, a Regulation on Public Instruction was carried out with the aim of organizing the entire system of education, including the elementary schools. Courses, examinations, teachers and finances were all taken into consideration in this series of decisions, which defined the second attempt to found the *Darülfünûn*. The university was officially named the *Darülfünûn-i Osmani*, and its internal organization was strongly influenced by the French model. According to the Regulation, it was to be divided into three departments : philosophy-literature, natural sciences-mathematics, and law. Required for graduation were three years of coursework and one year for the preparation of a thesis, totaling four years. Students sixteen years of age and in possession of a preparatory degree or its equivalent could enroll. Each department had a detailed curriculum culminating in a graduation thesis based on the student's own research. Teaching certificates were also granted. A museum, a library and a laboratory were to be opened at the *Darülfünûn*. The courses to be offered were inspired by their French counterparts. Arabic and Persian as eastern languages, and French, Greek, and Latin as Western languages would be offered in the department of philosophy and literature. Both Islamic and Roman law were to be taught in the department of law. The teaching of disciplines springing from both Islamic and Western cultures in these two departments reflects the desire of Ottoman intellectuals of the *Tanzimat* period to achieve a new " Ottoman cultural synthesis ".

Registration at the *Darülfünûn* began in 1869. An examination was administered to some one thousand applicants, of whom only 450 were accepted. The *Darülfünûn* opened with a grand ceremony, with Grand Vizier Ali Pasha, Minister of Education Safvet Pasha, and other dignitaries were present. Tahsin Efendi (Hodja Tahsin) was appointed Director of the *Darülfünûn*. He was appointed because it was believed that he could establish a harmonious balance between Islam and the West, between the old and the new. Indeed, Tahsin Efendi, having been educated in the traditional *medrese* system, then later taking on the duties of director at the Ottoman school in Paris, was well situated to find common ground between Islamic and Western cultures. Rather than stumbling on the contradictions between the two, Tahsin Efendi would put forward a new convergent view that would satisfy the *Tanzimat* dignitaries, who were in search of a synthesis between the two cultures.

However, the conditions were not yet conducive to opening the university for classes, due to a lack of funds, teachers, textbooks, and so on. Under these conditions, the regulations of the *Darülfünûn-i Osmanî* could not be fully implemented, and the programs of education in the three departments could not be realized as envisaged. Thus the same curriculum was followed in every department, and all students attended the same lessons. For these reasons the second attempt to establish the university did not bear the expected fruit.

In 1873, Minister of Education Safvet Pasha gave the task of setting up the *Darülfünûn* to Sawas Pasha, an Ottoman Greek who was then the Director of the *Mekteb-i Sultânî* (Imperial Lycée of Galatasaray). The condition under which Sawas Pasha had to work was that the third attempt at the *Darülfünûn* should not be a burden on the state treasury. The idea was to establish the *Darülfünûn* on the foundations of the *Mekteb-i Sultânî*, which had been functioning since 1868. Thus, an institution of higher education was founded based on an institution of secondary education. The new *Darülfünûn* would be called *Darülfünûn-i Sultânî*, and would consist of schools of law, sciences, and the arts. In official accounts, these three departments were referred to as the " Higher Schools " (*Mekâtib-i Aliye*).

The *Darülfünûn-i Sultânî*, which opened for classes in the 1874-1875 academic year, consisted of the schools of Letters, Law, and — instead of the originally intended school of science — Civil Engineering (*Mühendisin-i Mülkiye Mektebi*). During the first academic year, the name of *Mühendisin-i Mülkiye Mektebi* was changed to *Turuk u Maabir Mektebi* (School of Public Works, the Turkish translation of the French *L'Ecole des Ponts et Chaussées*).

Students who studied in the *Darülfünûn-i Sultânî* for four years would prepare and defend a scholarly thesis and graduate with the title of " Doctor ". Law students would be employed by the Ministry of Justice and engineers by the Ministry of Public Works. Graduates of the Faculty of Letters would be employed as teachers of literature. Students who did not prepare a dissertation would take an examination designed to be easier than the doctoral examination. Such non-dissertation graduates of the faculties of Letters, Law, and Public Works would be appointed as junior teachers, secondary lawyers, and engine drivers.

During the 1874-1875 academic year, the numbers of students who attended the schools of Law and Public Works were twenty-one and twenty-six respectively. All succeeded in the examinations that were given at the end of the academic year. We have no known documents telling us whether the School of Letters had started giving instruction by this time or not. In 1881, the schools of Law and Civil Engineering were attached to the Ministries of Justice and Public Works respectively, and continued to function successfully as independent schools.

At the end of the nineteenth century, the foundation of new schools of higher education where students could specialize in the fields of civil service, medicine, law, commerce, industry, engineering, and architecture were completed to answer the requirements of the State for qualified personnel.

However, on 14 February, 1895, the Grand Vizier Said Pasha submitted a petition to Sultan Abdülhamid II requesting that in addition to higher institutions where professional training was given, a *darülfünûn* should be established on the model of American and European universities, with five faculties (*darülicâze*). Here the students would receive scholarly education, not merely for the sake of becoming civil servants of the state bureaucracy, but for the pursuit of scholarly academic study in various fields.

After three failed attempts, the beginning of the twentieth century saw the successful foundation of this new *Darülfünûn* in 1900. This new institution would house several departments in a campus under one administration. Thus the university by the name of *Darülfünûn-i Shahane* was established. It was the product of fifty years' experience, and was further supported by a now-sufficient number of secondary schools, well-educated students, and firmly established faculties of law and medicine and other institutions of higher learning. The *Darülfünûn-i Shahane* was the first, most important, and most influential of all modern conventional Turkish universities.

Learned Societies

In the classical period, informal gatherings of scholars, poets, and intellectuals were part of the Ottoman cultural life. Such groups usually met in the mansions of distinguished persons and the learned to discuss issues of intellectual and scholarly interest. However, these informal meetings never led to the establishment of formal societies. Societies similar to those in Europe were started only in the mid-nineteenth century, following the *Tanzimat* reforms in the fields of education and culture. From then on, many learned and professional societies were founded by Ottomans to advance the goal of professional solidarity and learning among their members.

The first attempt at such a society was made in 1851 by high level statesmen who founded the *Encümen-i Danish* (Learned Society). This society, the first of its kind in the Islamic world, was in some respects similar to the Paris *Académie Française*. It became the model for Ottoman learned societies, and was soon followed by others founded by the foreigners living in Istanbul. The first of these was the *Société Orientale de Constantinople* founded in 1852. This was followed by the *Société de Médicine de Constantinople*, established by foreign physicians from allied countries who gathered in Istanbul during the Crimean war. This society, which received the title " Shahane " (Imperial) by the Sultan's decree, survived until the end of the Empire, and continues today under the name of *Türk Tip Dernegi* (Turkish Medical Society).

In 1861, a group of Ottoman intellectuals under the leadership of Münif Pasha established the *Cemiyet-i Ilmiye-i Osmaniye*, which would function in accordance with the State's educational and cultural policy. Its monthly *Mecmua-i Fünûn,* which contained news of modern European culture and science, was the first of its kind published by the Ottomans. The activities of this society were discontinued due to lack of funds.

Approximately ten years later, Hodja Tahsin Efendi and his friends founded the *Cemiyet-i Ilmiye*, with the objective of reviving notion of the Islamic-European synthesis. Though this organization only existed for a year, between1879-1880, it published a journal entitled *Mecmua-i Ulum,* of which only seven issues came out.

With the proclamation of the second *Meshrutiyet* (constitutional government) in 1908, chemists, engineers, architects, dentists, veterinary surgeons, pharmacists, and agriculturists all established their own societies to build solidarity and pursue professional interests. During the nineteenth century, the societies established to strengthen the solidarity among physicians and pharmacists were of long duration and proved very influential. However, learned societies founded for the introduction and dissemination of modern culture were small in number and soon petered out.

New Scientific Establishments

Parallelling expanding education in modern sciences such as medicine, astronomy, botany, zoology in the nineteenth century, a growing need was felt for institutions where these sciences were applied. The Department of Quarantine established in 1831 was one of the first institutions to implement modern medical practices — which they did particularly with regard to the annual pilgrimage to Mecca. In 1862, quarantine hospitals were built in Istanbul and other important cities along the roads to Mecca in Anatolia, Rumelia, and the Arab provinces to prevent the spread of infectious diseases. The Ottomans, who at this time were particularly concerned with public health, closely followed developments in Europe in the fields of microbiology and vaccination. When Pasteur discovered the rabies vaccine in 1885, a lecture was given in Istanbul on the subject, and a delegation of physicians was sent to Paris to learn more. They presented Pasteur with the official decoration of the Ottoman Empire together with a prize of 10,000 FF from Sultan Abdul Hamid II. After the delegation's return, with the collaboration of Ottoman and European physicians, a rabies laboratory (*Da'ul-Kelb Ameliyathânesi*) was founded. Later, a bacteriology laboratory was established to combat cholera epidemics. These institutions played an important role in the prevention of rabies, cholera, and small pox epidemics by promoting vaccination.

Another modern scientific institution established in the second half of the nineteenth century was the *Rasathâne-i Amire* (The Imperial Observatory). Affiliated to the Ministry of Education, the Observatory was built on European models and directed by the Frenchman M. Coumbary. Though called " the observatory ", this structure functioned more as a meteorology station than a place for making astronomical observations. Under M. Coumbary's direction, it sent out weather reports from the big cities and exchanged reports with stations in Europe. This office continued to function under Turkish administrators after the departure of Coumbary. In 1910, after the appointment of Fatin Efendi [Gökmen] (who was first educated at the *medrese* then at the *Darülfünûn* Faculty of Science) as Director, the observatory started making astronomical observations as well as preparing weather reports. The observatory became known as the *Kandilli Rasathânesi* (Kandilli Observatory) after it was moved to the district of Kandilli in Istanbul. Today, it is attached to Bogaziçi University and continues to carry out astronomical observations and seismic measurements.

Adoption of the metric system

The enterprising attitude of the Ottomans that we have seen in many other areas was again encountered in the field of metrology and standardization. With a law issued in 1869, the Ottoman State adopted the meter, gram and liter as the official units of length, weight and volume. Following the passage of the law, various measures were taken to facilitate the acceptance, use and dissemination of the metric system. Thus began a gradual transition from Ottoman to metric weights and measures. Due to the longstanding familiarity of the general public with the traditional system, and the abuse of the shopkeepers, the compulsory use of the metric system was posponed at regular intervals. However, the new system was regulary used in government offices and by certain professionals such as pharmacists and physicians. Greenwich mean time was later introduced to complete metrological integration of the Ottoman Empire and European countries. Following its adoption by communication and transportation companies and military and government offices in 1910s, mean time also came to use in daily life. The use of GMT was made obligatory with a law issued in 1926. Thus, as with the Ottoman and metric weights and measures, *alla turca* and *alla franca* times were used concurrently in the Ottoman Empire for many years.

The Emergence of Modern Turkish Scientific Literature

The nineteenth century scientific literature clearly shows that modern science co-existed with old Turko-Islamic scientific traditions in a number of fields. Examples can be found in works where geocentric and heliocentric systems of the universe were introduced together. Similar cases can be traced in medical works. Eighteenth-century Ottoman medical works included practical medical knowledge taken from Europe alongside old concepts such as the concept of four bodily humors (*ahlat-i arba'a*).

Toward the end of the eighteenth century, teachers at the Imperial School of Engineering, which had been established to teach modern sciences to military officers, started to translate or compile books from European scientific literature. Generally, the instructors benefited from the textbooks that were used in the European military technical schools. At the turn of the century, among the first scientific publications in the Ottoman Empire were about ten books on mathematics, geography, and engineering compiled and translated by Hüseyin Rifki Tamanî (d. 1817). Rifki Tamani's student — and later his successor as the chief instructor of the *Mühendishane* — Ishak Efendi (d. 1836) published thirteen volumes based on Western (particularly French) sources. Among these, *Mecmua-i Ulum-i Riyaziye* (Compendium of Mathematical Sciences), in four volumes, is of special importance, as it is the first attempt to prepare a comprehensive textbook on various sciences written in the language of any Muslim nation. The text treated the subjects of mathematics, physics, chemistry, astronomy, biology, botany, and mineralogy. Ishak Efendi's efforts to find Turkish equivalents for new scientific terms, and his role in disseminating information on modern sciences extended beyond the borders of Ottoman Turkey.

Instructors at the *Mühendishane* and graduates and instructors at the *Mekteb-i Tibbiye-i Shahane* (Imperial School of Medical Sciences), which was reformed and reopened in 1838, also contributed to the translation of European scientific books. As modern education became widespread and civilian education was reorganized after the proclamation of the *Tanzimat* (1839), new scientific and technical books were printed. By the middle of the nineteenth century, the number of printed books on modern science and technology, as well as the variety of subjects they addressed, had increased considerably. During the pre-*Tanzimat* period, between the establishment of the first Turkish printing press in 1727 and the proclamation of the *Tanzimat* in 1839, some thirty books on science were printed ; that figure reached approximately 250 during the *Tanzimat* period (1840-1876). There was an increase in the number of books that were printed in the field of mathematics and medicine, but a decrease in the fields of geography, military sciences, engineering, astronomy, and navigation.

After the *Tanzimat,* different subjects were taken on. For example, Dervish Pasha published the first book on chemistry in Turkish, titled *Usul-i Kimya* (*Elements of Chemistry*, Istanbul, 1848). The first book on modern zoology and botany, titled *Ilm-i Hayvanat ve Nebatat* (*Zoology and Botany*, Istanbul, 1865), was translated from the French by the Chief Physician Salih Efendi.

During the first three decades of the *Tanzimat* period, about four books were printed every year, but in the following seven years (1870-1876), this number went up to about eighteen books a year. This dramatic increase is indicative of the increasing interest in modern sciences in Ottoman society.

The adoption of Turkish as the language of instruction in the Imperial School of Medicine was instrumental in developing a Turkish medical literature. This resulted in the translation and publication of several books on medicine, particularly after 1870. The first of these was *Lugat-i Tibbiye* (*Medical Dictionary*, Istanbul, 1873).

The decrease noted above in the number books published in geography, engineering, and military sciences indicated a shift of interest from military to civilian areas. This shift is confirmed if one examines the prefaces of the books compiled on the same subjects in the nineteenth century before and after the *Tanzimat*. In his work *Mecmua-i Ulum-i Riyaziye,* Ishak Efendi mentions the importance of chemistry in the war industry. On the other hand, Kirimli Aziz Bey, in his work *Kimya-i Tibbi* (*Medical Chemistry,* Istanbul, 1868-1871) pointed out that chemistry, like medicine, was the basis of several industries and technologies of non-military character.

Some nineteenth-century books in the field of mathematics were original contributions rather than translations. Among them *Linear Algebra* written in English by Vidinli Hüseyin Tevfik Pasha is noteworthy as a valuable contribution to the development of linear algebra. This book, which is concerned with three-dimensional linear algebra and its application to geometry, was printed twice in Istanbul between 1882 and 1892.

During the eighteenth and nineteenth centuries, the primary languages of Ottoman scientific literature were Turkish and Arabic. Although there were a few works in Persian, these were rather rare, representing less than 1% of production. However, the relative share of Arabic and Turkish works varied between manuscripts and printed works. While almost all the printed books about modern science and technology produced during these two centuries in Istanbul were in Turkish, a significant number of manuscripts were still written in Arabic.

During the eighteenth century, the total number of books on astronomy produced in the Ottoman world, both in manuscript and print form, were 331. Of these, 221 were written in Arabic, 101 in Turkish, 2 in Persian and the remaining 7 in a mixture of these three languages. In the nineteenth century, the total number was 263, of which 137 were written in Arabic, 123 in Turkish, and 3 in the mixed languages. These figures clearly show the increase in the use of Turkish — even in the context of a decrease in the total number of books. This shift is also indicative of the increase in the number of schools and educational institutions during the nineteenth century, and the promptness with which the resulting transition from the Eastern manuscript tradition to the modern printed book was made.

Most of the astronomy books written in the Ottoman Empire outside the Arabic-speaking lands were in Turkish. Although we do not have statistical figures for other sciences, we may assume that the same pattern holds for them. We may thus conclude that in the eighteenth and nineteenth century scientific manuscripts were produced in both in Arabic and Turkish, while most of the increasing number of printed works were in Turkish.

Our information on research-oriented activities at Ottoman institutions is limited. However, in the second decade of the twentieth century, research activities were included in the objectives of the *Darülfünûn*.

In the second half of the nineteenth century there are some significant — yet not well documented — research-driven works produced in Europe by Ottoman scholars who pursued their studies abroad. After these scholars' return from Europe, they took up positions in the modern educational institutions founded all over the Ottoman Empire.

At the beginning of the twentieth century, Ottoman Turkish was a well-developed scientific language with an elaborate terminology that was used to write about many modern sciences. There were sufficient textbooks and, to a lesser degree, authentic publications in Ottoman. This was a long journey from the fourteenth century, when the first scientific Turkish literature was produced by the Ottoman Turks who settled in Anatolia.

Conclusion

In this paper, scientific developments and changes in the Ottoman State over the course of six centuries have been examined, while scientific activities in both institutional and private settings have been studied from different viewpoints. This paper, which may be considered as a summary of the history of Ottoman science, divides Ottoman science into two eras. First, the classical Ottoman science that was established and developed under the influence of pre-Ottoman Islamic traditions, and second, the Western scientific tradition who expanded with the European political and military influence. In reviewing these two eras, the decisive influence exerted by pre-Ottoman Islamic scientific traditions on Ottoman scientific life is emphasized. However, instead of just statically imitating their classical inheritance, the Ottomans made original and new contributions to the classical Islamic science in many areas, ultimately carrying the tradition to a higher level.

The Ottoman sultans, while establishing new *medreses* across the vast Empire in a very short span of time, allowed existing *medreses* to continue to function. During the first centuries of the Empire, scholars who came from other Islamic lands were employed as instructors in Ottoman *medreses*. Soon, the *medreses* improved their programs of instruction by restructuring curricula, setting up new individual systems and organizing their staff. We must note that the attention and patronage given to the rational sciences during Sultan Mehmed II the Conqueror's reign was key to these subjects' subsequent inclusion in the traditional *medrese* curriculum. The Süleymaniye Complex, founded by Sultan Süleyman the Magnificent, was the last and best link in the development of the classical *medrese* system. The fact that *Daruttip* which was established within this complex was completely devoted to medical education is particularly noteworthy.

In addition to the *medreses*, scientific education was also given in institutions such as the *muvakkithane* and *shifahane*, where applied medicine, mathematics, and astronomy were practiced. Both traditional and new instruments of astronomy were used together In the Istanbul Observatory, which was founded towards the end of the sixteenth century by Taqi al-Din Rasid. Here, Islamic astronomical observations and studies continued. Short but original studies were also conducted and immediately published as books. The classical Ottoman scientific literature was generally produced within the *medrese* milieu. Numerous books were written by authors in Arabic and Turkish, and in lesser numbers in Persian. These books were primarily on religious subjects, but also included efforts in the fields of mathematics, medicine, and astronomy.

One of our most important observations is that books written by scholars of Muslim or Jewish origin who came from non-Ottoman lands (including, importantly, Andalusia) have a significant place in Ottoman scientific literature. When seen within the framework of classical Ottoman science, the patronage given by sultans and statesmen to *medreses* and scholars, and their evaluation of original works are all brought into focus. The Ottomans, who were aware of scientific and geographical advances in Europe, were beginning to take a keen interest in Western science and technology, particularly towards the end of the seventeenth century. In order to learn new technologies of warfare, they switched from their old policy of limited, selective adoption of scientific advances to one of more intensive borrowing and imitation.

At the beginning of the eighteenth century, the Ottomans became more interested in European science and technology. With the assistance of European experts, important steps were taken in transforming the army. After concentrated efforts to introduce modern military and technical education, new institutions were established. This started with the establishment of the *Ulufeli Humbaraci Ocagi* and continued with the opening of the *Mühendishane* and other military schools in the nineteenth century. Among the military schools, institutions of modern medicine took pride of place. In addition to the military and technical schools, modern civilian schools were also founded, thus spreading the modernization of scientific education among a wider segment of the population. To fill in the teaching staff positions in these institutions, a large number of students were sent to Europe for training. One of our most important observations is that the *Darülfünûn-i Shahane*, which was finally opened in 1900 after three unsuccessful attempts, was the pioneer in the foundation of the present-day schools of higher education in Turkey.

In addition to the official institutions of education, Ottoman intellectuals founded numerous professional and learned societies and made great efforts in the production of a modern Turkish scientific literature. They published a large number of books, mainly translations, and including some dictionaries. The majority of the works published in the eighteenth and nineteenth centuries were in Turkish, which indicates that Turkish had by then been accepted as a scientific language.

Though European science and technology was followed closely by the Ottomans, the modern scientific approach was never wholly accepted. Notably, the understanding of science as based on research was lacking. Due to the lack of this fundamental understanding, Ottomans were not able to conduct systematic research studies parallel to those seen emerging in Russia and Japan. Yet in spite of this setback, Ottoman scholars and intellectuals were able to produce many scholarly works in the sciences, and continued to develop and work on Ottoman Turkish as a scientific language. At the beginning of the twentieth century, ideas in a wide range of scientific subjects could easily be expressed in Ottoman Turkish.

The cultural and scientific heritage that was accumulated during the Ottoman period constitutes the scientific and cultural background of the numerous successor states to that Empire in the Balkans and the Middle East, but particularly that of the Republic of Turkey.

BIBLIOGRAPHY

A. Adivar, *Osmanli Türklerinde Ilim,* 5th ed., ed. A. Kazancigil and S. Tekeli. Istanbul, Remzi Kitabevi, 1991.

S. Aydüz, *Osmanli Devleti'nde Müneccimbasilik ve Müneccimbasilar*, Unpublished M.A. thesis, Istanbul University, Faculty of Letters, Dept. of the History of Science, 1993 (supervisor : E. Ihsanoglu).

S. Aydüz, " *Osmanli Devleti'nde Müneccimbasilik* ", F. Günergun (ed.), *Osmanli Bilimi Arastirmalari II (Studies in Ottoman Science II)*, Istanbul, Istanbul University, 1998, 159-207.

S. Çavusoglu, *The Kadizadeli Movement : An Attempt of Sheri'at-Minded Reform in the Ottoman Empire*, Unpublished Ph.D. diss., Princeton University, Near Eastern Studies Department, 1990 (Supervisor : C. Kafadar).

T.C. Goodrich, *The Ottoman Turks and the New World, A Study of Tarih-i Hind-i Garbi and Sixteenth-Century Ottoman Americana*, Wiesbaden, Otto Harrassowitz, 1990.

F. Günergun, " Du Zira au Mètre : Une Transformation Métrologique dans l'Empire Ottoman ", Patrick Petitjean, C. Jami and A.-M. Moulin (eds), *Science and Empires*, Dordrecht, Boston, London, Kluwer Academic Publishers, 1992, 103-110 (*Boston Studies in the Philosophy of Science*, vol. 136).

F. Günergun, " Metric System in Turkey : Transition Period (1881-1934) ", *Journal of the Japan-Netherlands Institute*, vol. 6 (W.G.J. Remmelink (ed.), Papers of the Third Conference on the Transfer of Science and Technology between Europe and Asia since Vasco da Gama (1498-1998)), Tokyo, 1996, 243-256.

F. Günergun, " Standardization in Ottoman Turkey ", F. Günergun and S. Kuriyama (eds), *Introduction of Modern Science and Technology to Turkey and Japan*, Kyoto, International Research Center for Japanese Studies, 1998, 205-225.

E. Ihsanoglu, *Bashoca Ishak Efendi : Türkiye'de Modern Bilimin Öncüsü* (*Chief Instructor Ishak Efendi : Pioneer of Modern Science in Turkey*), Ankara, Kültür Bakanligi, 1989.

E. Ihsanoglu, " Some Remarks on Ottoman Science and its Relation with European Science & Technology up to the end of the Eighteenth Century ", *Journal of the Japan-Netherlands Institute,* vol. 3 (W.G.J. Remmelink (ed.),

Papers of the First Conference on the Transfer of Science and Technology between Europe and Asia since Vasco da Gama (1498-1998), Tokyo 1991, 45-73.

E. Ihsanoglu, " Ottoman Science in the Classical Period and Early Contacts with European Science and Technology ", E. Ihsanoglu (ed.), *Transfer of Modern Science & Technology to the Muslim World*, Istanbul, The Research Centre for Islamic History, Art and Culture, 1992, 1-48.

E. Ihsanoglu, " Introduction of Western Science to the Ottoman World : A Case Study of Modern Astronomy (1660-1860) ", E. Ihsanoglu (ed.), *Transfer of Modern Science & Technology to the Muslim World*, Istanbul, The Research Centre for Islamic History, Art and Culture, 1992, 67-120.

E. Ihsanoglu, " Ottomans and European Science ", P. Petitjean, C. Jami and A.-M. Moulin (eds), *Science and Empires*, Dordrecht, Boston, London, Kluwer Academic Publishers, 1992, 37-48 (*Boston Studies in the Philosophy of Science*, vol. 136).

E. Ihsanoglu, " Tanzimat Öncesi ve Tanzimat Dönemi Osmanli Bilim ve Egitim Anlayishi ", *150. Yilinda Tanzimat,* ed. Hakki Dursun Yildiz, Ankara, Türk Tarih Kurumu, 1992, 335-395.

E. Ihsanoglu, " Bashoca Ishak Efendi, Pioneer of Modern Science in Turkey ", *Decision Making and Change in The Ottoman Empire*, Missouri, The Thomas Jefferson University Press at Northeast Missouri State University, 1993, 157-168.

E. Ihsanoglu, F. Günergun, " Tip Egitiminin Türkçelesmesi Meselesinde Bazi Tesbitler ", A. Terzioglu (ed.), *Türk Tip Tarihi Yilligi-Acta Turcica Historiae Medicinae I*, Istanbul, 1994, 127-134.

E. Ihsanoglu, " Ottoman Science ", H. Selin (ed.), *Encyclopaedia of the History of Science, Technology, and Medicine in Non-Western Cultures*, London, Kluwer Academic Publishers, 1997, 799-805.

E. Ihsanoglu, R. Sesen, *et al.*, *Osmanli Astronomi Literatürü Tarihi (History of Astronomy Literature During the Ottoman Period)*, 2 vols, ed. and foreword E. Ihsanoglu, Istanbul, The Research Centre for Islamic History, Art and Culture, 1997.

E. Ihsanoglu, " Modernization Efforts in Science, Technology and Industry in the Ottoman Empire (18th and 19th Centuries) ", F. Günergun, S. Kuriyama (eds), *The Introduction of Modern Science and Technology to Turkey and Japan*, Kyoto, International Research Center for Japanese Studies, 1998, 15-35.

E. Ihsanoglu, " Changes in Ottoman Educational Life and Efforts towards Modernization in the 18th and 19th Centuries ", F. Günergun, S. Kuriyama (eds), *The Introduction of Modern Science and Technology to Turkey and Japan*, Kyoto, International Research Center for Japanese Studies, 1998, 119-136.

E. Ihsanoglu, " Osmanli Egitim ve Bilim Müesseleri ", E. Ihsanoglu (ed.), *Osmanli Devleti ve Medeniyeti Tarihi (History of the Ottoman State and Civilisation)*, vol. 2, Istanbul, The Research Centre for Islamic History, Art and Culture, 1998, 223-359.

E. Ihsanoglu, " Osmanli Bilimi Literatürü ", E. Ihsanoglu (ed.), *Osmanli Devleti ve Medeniyeti Tarihi (History of the Ottoman State and Civilisation),* vol. 2, Istanbul, The Research Centre for Islamic History, Art and Culture, 1998, 363-444.

H. Inalcik, *The Ottoman Empire : The Classical Age, 1300-1600,* Trans. N. Itzkowitz and C. Imber, London, Weidenfeld and Nicolson, 1973.

M. Kaçar, " Osmanli Imparatorlugunda Askeri Sahada Yenilesme Döneminin Baslangici ", F. Günergun (ed.), *Osmanli Bilimi Arastirmalari (Studies in Ottoman Science)*, Istanbul, Istanbul University, 1995, 227-238.

M. Kaçar, *Osmanli Devletinde Mühendishanelerin Kurulusu ve Bilim ve Egitim Anlayisindaki Degismeler* (*The Foundation of " Mühendishane" in the Ottoman State and the Changes in the Scientific and Educational Life*), Unpublished Ph.D. diss. Istanbul University, Faculty of Letters, Dept. of History of Science, 1996 (supervisor : E. Ihsanoglu).

M. Kaçar, " Osmanli Imparatorlugu'nda Askeri Teknik Egitimde Modernlesme Çalismalari ve Mühendishanelerin Kurulusu (1808'e kadar)", F. Günergun, *Osmanli Bilimi Arastirmalari II* (*Studies in Ottoman Science II*), Istanbul, Istanbul University, 1998, 69-137.

E. Kâhya, *El-'Itaqi, The Treatise on Anatomy of Human Body*, Islamabad, National Hijra Council, One Hundred Great Books of Islamic Civilization series n° 85 (a), 1990 (Original Turkish text with English translation).

I. Miroglu, " Istanbul Rasathanesine Ait Belgeler ", *Tarih Enstitüsü Dergisi*, vol. 3, Istanbul University, Faculty of Letters, 1973, 75-82.

G. Russell, " The Owl and the Pussycat : The Process of Cultural Transmission in Anatomical Illustration ", E. Ihsanoglu (ed.), *Transfer of Modern Science & Technology to the Muslim World*, Istanbul, The Research Centre for Islamic, History, Art and Culture, 1992, 180-212.

N. Sari, Z. Bedizel, " The Paracelsusian Influence on Ottoman Medicine in the Seventeenth and Eighteenth Centuries ", E. Ihsanoglu (ed.), *Transfer of Modern Science & Technology to the Muslim World*, Istanbul, The Research Centre for Islamic, History, Art and Culture, 1992, 157-179.

A. Sayili, *The Observatory in Islam*. Ankara, Turkish Historical Society, 1960.

S. Sherefeddin, *Cerrahiyyetü'l-Haniyye,* 2 vols, ed. I. Uzel, Ankara, Turkish Historical Society, 1992.

S. Tekeli, " Onaltinci Yüzyil Trigonometri Çalismalari Üzerine bir Arastirma, Copernicus ve Takiyüddin (Trigonometry in the Sixteenth Century, Copernicus and Taqi al Din) ", *Erdem,* vol. 2/4, Ankara 1986, 219-272.

S. Tekeli, " Osmanlilarin Astronomi Tarihindeki En Önemli Yüzyili ", *Fatih'ten Günümüze Astronomi,* Prof. Dr. Nüzhet Gökdogan Sempozyumu, Istanbul Üniversitesi'nin Kurulusunun 540. Yildönümü, Istanbul University, 1994, 69-85.

S. Tekeli, " Taqi al-Din ", H. Selin (ed.), *Encyclopaedia of the History of Science, Technology, and Medicine in Non-Western Cultures*, Dordrecht, Boston, London, Kluwer Academic Publishers, 1997, 934-935.

A. Terzioglu, *Moses Hamons Kompendium der Zahnheilkunde aus dem Anfang des 16. Jahrhunderts*, München, 1977.

A. Terzioglu, " Bîmâristan, Islâm dünyasinda klasik hastahanelerin genel adi ", *Türkiye Diyanet Vakfi Islâm Ansiklopedisi,* vol. 6, Istanbul, 1992, 163-178.

I. Uzel, " Dentistry in the Early Turkish Medical Manuscripts ", Ph.D. diss. Istanbul University, Istanbul Medical Faculty, 1979, (supervisor : A. Terzioglu).

A.S. Ünver, " Osmanli Türkleri Ilim Tarihinde Muvakkithaneler ", *Atatürk Konferanslari 1971-72*, vol. 5, 1975, 217-257.

T. Zorlu, *Süleymaniye Tip Medresesi*, Unpublished M.A. thesis, Istanbul University, Faculty of Letters, Dept. of History of Science, 1998 (supervisor : E. Ihsanoglu).

LES ACTIVITÉS MATHÉMATIQUES AU MAGHREB À L'ÉPOQUE OTTOMANE (XVIe-XIXe SIÈCLES)

Ahmed DJEBBAR

INTRODUCTION

La longue période de l'histoire du Maghreb qui commence dans les premières décennies du XVe siècle, avec les initiatives du corsaire turc ᶜArrūj (m. 1518) et de son frère Khayr ad-Dīn (m. 1546), et qui s'achève en 1830 avec le débarquement des troupes du roi de France Charles X (1824-1830), est indissociable de celle encore plus longue de l'Empire ottoman. D'une manière plus précise, elle s'inscrit dans l'histoire des relations multiformes qui se sont établies entre, d'une part, cet Empire et les différents pouvoirs européens qui se sont opposés à lui pour le contrôle de la Méditerranée occidentale et, d'autre part, entre le pouvoir central ottoman et les différentes forces qui agissaient dans sa périphérie (pouvoirs locaux, tribus, confréries religieuses, *etc*.).

Dans ce cadre, les parties centrale et orientale du Maghreb ont connu, sur les plans économique et politique, des évolutions importantes et des changements profonds. Mais, lorsqu'on se place sur le plan culturel et scientifique, on ne constate pas de ruptures significatives. On a même l'impression, au vu des sources aujourd'hui accessibles, que c'est plutôt la continuité qui a dominé, tant au niveau des grandes tendances héritées des XIVe-XVe siècles qu'au niveau des activités et de la production des hommes de sciences, et plus particulièrement de ceux qui se sont occupés de mathématiques et d'astronomie.

Dans cette communication, nous nous proposons d'illustrer ce phénomène de continuité et de donner quelques éléments qui pourraient aider à mieux cerner sa nature, sans toutefois prétendre l'expliquer. Pour cela, nous présenterons les éléments connus concernant les activités mathématiques, au sens large, qui ont eu lieu dans la partie du Maghreb qui a été intégrée à l'Empire ottoman.

Pour mieux faire ressortir les éléments de continuité, à la fois au niveau de l'espace maghrébin dans son ensemble et au niveau de son histoire scientifique, nous serons amenés à aborder deux aspects complémentaires à notre sujet.

Le premier concerne les éléments caractéristiques des mathématiques qui ont été produites aux XIVe-XVe siècles, au Maghreb dans son ensemble, et la situation qui prévalait dans ce domaine à la veille de l'avènement du pouvoir ottoman en Méditerranée occidentale.

Le second aspect concerne les activités mathématiques qui se sont poursuivies dans la partie du Maghreb non intégrée à la mouvance ottomane, c'est-à-dire le Maghreb Extrême dont la gestion politique était assurée par la dynastie sacdide puis, à partir de 1631, par la dynastie calawite. L'évocation de ce point est indispensable, compte tenu des relations qui se sont poursuivies entre les foyers culturels et scientifiques des deux régions du Maghreb, durant les trois siècles de présence ottomane[1].

Par ailleurs, comme il ne nous est pas possible d'évoquer dans le détail la production scientifique de cette période, nous nous contenterons de donner des titres d'ouvrages avec, lorsque le propos l'exige, une description succincte de leurs contenus, en attendant de publier un travail de biobibliographie, en préparation, sur les activités mathématiques et astronomiques au Maghreb[2].

Le contexte politique et socio-culturel

Comme dans les autres régions de l'Empire musulman, l'intervention ottomane au Maghreb (qui a eu lieu exactement à la même période) ne s'est pas heurtée à une grande résistance. Et là où cette dernière s'est manifestée, comme en Ifriqya par exemple, elle a été beaucoup plus la conséquence des interventions ou de la présence de l'Espagne (et dans une moindre mesure du Portugal) que l'expression d'un refus du nouveau pouvoir de la part de la majorité des Maghrébins et de leurs dirigeants.

Sans vouloir minimiser les facteurs internes, il semble que c'est cet antagonisme ibéro ottoman et ce qu'il symbolisait alors dans l'esprit des contemporains qui est l'élément essentiel dans l'analyse de l'intervention ottomane au Maghreb et des conséquences de cette intervention sur les plans politique, économique et culturel.

Quant aux liens politiques entre le pouvoir central ottoman et les provinces maghrébines, tout au long de ces trois siècles, on constate qu'ils n'ont cessé d'évoluer pour aboutir à des formes d'autonomie très larges s'exerçant dans le cadre d'une suzeraineté toute théorique. C'est bien ce que l'on observe

1. Sur certains aspects de l'activité scientifique sous les dynasties sacadite et calawite, *cf.* M. Al-Mānūnī, " Asātidhat al-handasa wa mu'allifūhā fī l-Maghrib as-sacdī " [Les professeurs de géométrie dans le Maghreb sacadide], *Revue Dacwat al-ḥaqq*, n° 2 (Rabat, 9e année, 1965), 101-104 ; Ibn Zaydān, *an-Nahḍa al-cilmiyya cala cahd ad-dawla al-calawiyya* [La renaissance scientifique à l'époque de l'État calawite], Ms. Rabat Ḥasaniyya, n° 3182.

2. A. Djebbar, *Mathématiques et mathématiciens du Maghreb (IXe-XIXe siècles) : un essai biobibliographique* (en préparation).

lorsqu'on compare entre elles les trois grandes phases qui composent cette période.

La première phase (début XVIe-milieu XVIIe siècle) est celle de la cogestion du pouvoir entre les Pachas, représentant le pouvoir central ottoman, et l'Ojaq qui est la milice des janissaires, tous originaires d'Anatolie. La seconde est celle de l'Ojaq seul (milieu XVIIe-fin XVIIe siècle). La troisième est celle des Dey (fin XVIIe-milieu XIXe siècle) qui étaient en fait de véritables rois locaux sans autres liens avec la Sublime Porte que l'acte d'investiture et l'évocation du nom du sultan à l'occasion de la prière du vendredi.

Partant de ces constats pour lesquels il semble y avoir un certain consensus parmi les différents auteurs qui ont écrit sur l'histoire politique ou culturelle du Maghreb, on observe des clivages radicaux lorsqu'il s'agit d'interpréter l'antagonisme ibéro-ottoman des XVIe-XVIIIe siècles, et de caractériser la nature des pouvoirs qui se sont succédé au Maghreb jusqu'à l'intervention française de 1830.

C'est ainsi que certains auteurs considèrent l'intervention puis la présence ottomanes comme une conquête et une occupation[3]. D'autres y voient plutôt l'avènement des premières formes des trois Etats maghrébins d'Algérie, de Tunisie et de Libye[4]. D'autres encore interprètent les initiatives du pouvoir central ottoman ou de ses représentants au Maghreb comme une série d'interférences dans un processus interne plus profond et plus ancien qui aurait été à l'oeuvre depuis le XIVe siècle[5].

Ces différentes lectures sont importantes pour le sujet qui nous préoccupe ici dans la mesure où elles proposent aussi, implicitement ou explicitement, des explications différentes à propos du contenu et de la qualité des activités culturelles et scientifiques au Maghreb durant la phase ottomane de son histoire.

Cela dit, lorsqu'on quitte l'espace strictement politique pour ne s'intéresser qu'au contexte socioculturel dans lequel ont eu lieu les activités scientifiques, on constate qu'en plus des facteurs propres à telle ou telle région du Maghreb, qui étaient d'ailleurs à l'oeuvre bien avant le XVIe siècle, des phénomènes nouveaux font leur apparition et d'autres plus anciens sont réactivés, à la faveur des changements politiques locaux et régionaux.

Pour le sujet qui nous intéresse ici, trois phénomènes devraient, à notre avis, être privilégiés dans les recherches futures parce que les travaux qui leur ont

3. A. Sa^{c}dallah, *Tārīkh al-Jazā'ir ath-thaqāfī min al-qarn al-^{c}āshir ilā al-qarn ar-rābi^{c} ^{c}ashar al-hijrī* [Histoire culturelle de l'Algérie, du dixième au quatorzième siècle de l'Hégire], 2e éd., Alger, al-Mu'assasa al-waṭaniyya li l-kitāb, 1985.

4. A.T. Al-Madanī, *Ḥarb ath-thalath mi'at sana bayna l-Jazā'ir wa Isbānyā (1492-1792)* [La guerre de trois cents ans entre l'Algérie et l'Espagne (1492-1792)], Alger, Société Nationale d'Édition et de Diffusion, 1976.

5. A. Laroui, *L'histoire du Maghreb, un essai de synthèse,* Paris, Maspéro, 1970, 227-274.

été consacrés jusqu'à maintenant n'ont pas toujours pris en compte leurs éventuels effets sur l'activité scientifique de cette époque.

Le premier phénomène est interne aux sociétés maghrébines dans leur ensemble et il y constitue, depuis le XVe siècle, un élément incontournable, tant sur le plan politique que dans le domaine de la culture au sens large. Il s'agit du phénomène des zāwiyya avec tout ce qu'il a représenté comme foyers de contestation des pouvoirs locaux ou de résistance aux interventions ibériques, comme antagonisme entre les différentes confréries qui le représentaient (Darqāwā, Tijāniyya, Shādhiliyya, Qādiriyya, pour ne citer que les plus importantes) et, enfin, comme source d'inspiration pour une certaine production intellectuelle. Nous connaissons le rôle joué par ces zāwiyya, en particulier sur le plan politique, dans la mobilisation de la population contre la pénétration ibérique au Maghreb[6]. Mais leur rôle culturel n'a pas encore été évalué et comparé à celui des villes[7].

Le second phénomène est celui de l'émigration massive des derniers musulmans d'al-Andalus vers certaines villes du Maghreb Central et de l'Ifriqya. Cette émigration en plusieurs vagues, qui a commencé bien avant la chute de Grenade (1492), semble avoir culminé au début du XVIIe siècle avec le départ, en une seule fois, d'environ cent mille personnes. L'apport de ces populations a été d'abord un ensemble de savoir-faire qui s'est investi dans les secteurs administratifs, manufacturiers et agricoles. Mais, pour une partie d'entre elles, il s'agissait plutôt d'un savoir et d'une culture livresques qui vont être diffusés dans les milieux intellectuels du Maghreb[8].

Le troisième et dernier phénomène qui est, comme le second, extérieur à la région et porteur, comme lui, de nouveautés sur le plan socioculturel, englobe tous les éléments importés du centre de l'Empire ottoman. Cet apport a bien sûr concerné en premier lieu les secteurs administratifs, militaires et économiques. C'est d'ailleurs ce que les historiens évoquent le plus souvent. Mais il est difficile d'admettre qu'il n'ait pas concerné, même indirectement, les activités intellectuelles en général et les sciences en particulier. Nous évoquerons plus loin un de ses aspects relatif au rôle de la ville d'Istanbul dans le transfert des sciences européennes vers l'Empire et nous nous interrogerons sur les réactions à ce transfert de la part des foyers scientifiques de la périphérie. Quant

6. *Op. cit.*, 229-232.

7. Pour le cas du Maghreb Central, cf. la riche étude de A. Sa^c^dallah, " Tārīkh al-Jazā'ir ath-thaqāfī... ", *op. cit.*, vol. I, 464-527.

8. H.H. Abdalwahhab, " Coup d'œil général sur les rapports ethniques étrangers en Tunisie ", *Revue Tunisienne*, XXIV (1917), 305-316, 371-379 ; publié également dans *Cahiers de Tunisie,* XVIII (1970), 149-169 ; H.-J. Kress, " Andalusische Strukturelemente in der Kultur Geographischen Genese Tunisiens ", *Marburgger Geographische Schriften,* n° 83 (1977), 257-284. Traduction française : " Eléments structuraux andalous dans la genèse de la géographie culturelle de la Tunisie ", *IBLA,* n° 145 (1980-1), 3-45 ; M. De Epalza, " Nouveaux documents sur les Andalous en Tunisie au début du XVIIIe siècle ", *Revue d'Histoire Maghrébine,* n° 17-18 (1980), 79-108.

aux autres aspects, leur connaissance, qui dépend en particulier de l'étude des archives ottomanes et maghrébines non encore exploitées, devrait nous éclairer sur les différents comportements qui sont apparus face aux autres éléments de " modernité ".

Si nous avons rappelé ces aspects liés, directement ou indirectement, à la présence ottomane au Maghreb et pas d'autres, c'est parce qu'il nous a semblé qu'ils constituent l'essentiel des éléments qui caractérisent le contexte dans lequel ont été pratiquées les activités scientifiques que nous allons évoquer.

Mais, avant cela, il est nécessaire de présenter un aperçu très bref de ce que nous savons aujourd'hui sur les activités mathématiques et astronomiques au Maghreb avant le XVIe siècle.

Les mathématiques au Maghreb avant le XVIe siècle

Les premières activités mathématiques au Maghreb, dans le cadre de la civilisation arabo-musulmane, ont eu lieu en Ifriqya, d'abord à Kairouan, dès la fin du VIIIe siècle, puis peut-être à Tahert, et elles se sont poursuivies tout au long du IXe siècle. Pour le Xe siècle, qui correspond à l'avènement du pouvoir fatimide au Maghreb, les sources sont très pauvres en information sur les sciences. On y trouve quelques noms associés aux mathématiques, soit pour leur activité d'enseignement soit à cause de leur intérêt pour ces disciplines, comme ce fut le cas pour le calife fatimide al-Mui^czz[9]. Les informations sur les mathématiciens et les astronomes du XIe siècle sont plus nombreuses mais elles restent encore très fragmentaires. Elles sont toutefois suffisantes pour nous permettre de constater la prééminence, à cette époque, des foyers scientifiques d'Ifriqya[10].

Il faudra en fait attendre le XIIe siècle pour disposer d'informations relativement abondantes et précises sur le contenu de la production mathématique et sur ses auteurs. On constate d'ailleurs que, chez les mathématiciens maghrébins de l'époque ottomane, les références aux écrits de leurs prédécesseurs ne sont pas antérieures au XIIe siècle. Les auteurs de cette période dont on cite encore les écrits après le XVIe siècle, ou dont on évoque les noms, sont essentiellement al-Qurashi (m. 1184) pour son traité d'algèbre et, surtout, pour sa méthode de résolution des problèmes d'héritage[11], al-Ḥaṣṣār (XIIe s.) avec son

9. A. Djebbar, " Quelques éléments nouveaux sur les activités mathématiques arabes dans le Maghreb oriental (IXe-XVIe siècles) " dans *Actes du IIe Colloque Maghrébin d'Histoire des Mathématiques Arabes Tunis, 1-3 Décembre 1988* (Tunis, Université de Tunis I-I.S.E.F.C.-A.T.S.M., 1990), 61-63.

10. *Op. cit.*, 53-73.

11. M. Zerouki, " Abū l-Qāsim al-Qurashī : ḥayātuhū wa mu'llafātuhū ar-riyyāḍiyya " [Abū l-Qāsim al-Qurashī : sa vie et ses écrits mathématiques], *Cahier du Séminaire Ibn al-Haytham,* n° 5 (Alger, E.N.S. de Kouba, 1995), 10-19.

manuel *Kitāb al-bayān wa t-tadhkār*[12], et Ibn al-Yāsamīn (m. 1204) pour son poème algébrique[13]. Pour ce siècle, il y a aussi Ibn Munᶜim (m. 1228) dont l'enseignement a marqué les mathématiciens de la génération suivante, originaires du Maghreb Extrême, mais dont l'écrit principal, *Fiqh al-ḥisāb* [La science du calcul], ne semble pas avoir été utilisé au-delà du XVᵉ siècle[14].

Dans l'histoire des activités scientifiques au Maghreb, le XIVᵉ siècle constitue une étape importante et ce pour plusieurs raisons. En premier lieu, on observe un développement quantitatif de la production mathématique qui est confirmé à la fois par les témoignages des bibliographes et par les manuscrits qui nous sont parvenus. En second lieu, on constate que le contenu de cette production est le résultat d'une synthèse entre différentes traditions (orientale, andalouse et maghrébine). En effet, dans l'état actuel de nos connaissances, on peut dire que la production mathématique de ce siècle est, en grande partie, une reprise partielle de ce qui avait été déjà découvert ou assimilé au cours des siècles précédents, au Maghreb et ailleurs. Les contributions nouvelles sont en effet exceptionnelles ; ce qui ne peut que confirmer, pour les mathématiques, les conclusions auxquelles avait, déjà à son époque, abouti Ibn Khaldūn lorsqu'il a évoqué le déclin de certaines activités scientifiques au Maghreb et en Andalus[15].

Le mathématicien qui a joué un rôle clé dans la synthèse du XIVᵉ siècle est sans conteste Ibn al-Bannā (m. 1321). Il semble en effet avoir été l'initiateur d'une tradition qui s'est étendue aux différentes régions du Maghreb et qui a même atteint l'Egypte. Cette tradition est celle des commentaires. Il y eut ainsi, tout au long du XIVᵉ et du XVᵉ siècle plus de quinze ouvrages plus ou moins importants consacrés à l'explication ou au développement, et parfois même à la critique, du manuel d'Ibn al-Bannā, *Talkhīṣ aᶜmāl al-ḥisāb* [L'abrégé des opérations du calcul][16].

L'analyse détaillée des chapitres les plus importants de ces commentaires nous permet d'avancer quelques remarques qui permettent d'éclairer notre pro-

12. M. Zoubeidi, *Kitāb al-bayān wa t-tadhkār d'al-Ḥaṣṣār* [Le livre de la preuve et du rappel d'al-Ḥaṣṣār], Magister en Histoire des Mathématiques, Alger, E.N.S. de Kouba, Magister en préparation.

13. T. Zemouli, *Mu'allafāt Ibn al-Yāsamīn ar-riyāḍiyya* [Les écrits mathématiques d'Ibn al-Yāsamīn], Magister d'Histoire des Mathématiques, Alger, E.N.S. de Kouba, 1993.

14. A. Djebbar, *L'analyse combinatoire au Maghreb : l'exemple d'Ibn Munᶜim (XIIᵉ-XIIIᵉ siècles)*, Paris, Université de Paris-Sud, 1985 (*Publications Mathématiques d'Orsay*, n° 85-01) ; A. Djebbar, " Quelques aspects de l'activité mathématique dans le Maghreb Extrême (XIIᵉ-XVᵉ s) " (en Arabe), *IVᵉ Colloque Maghrébin sur l'Histoire des Mathématiques Arabes, Fez, 2-4 Décembre 1992* (à paraître).

15. ᶜA. Ibn Khaldūn, *Kitāb al-ᶜibar* [Le livre des leçons], Beyrouth, Dār al-kitāb al-lubnānī-Maktabat al-madrasa, 1983, vol. II, 893, 896.

16. M. Aballagh, A. Djebbar, *Ḥayāt wa mu'allafāt Ibn al-Bannā (maᶜa nuṣūṣ ghayr manshūra* [La vie et l'œuvre d'Ibn al-Bannā (avec des textes inédits)]. A paraître dans les Publications de la Faculté des Lettres et Sciences Humaines de Rabat.

pos sur les activités mathématiques au Maghreb à l'époque ottomane.

En premier lieu, on constate que les mathématiques qui y sont exposées ont gardé leur niveau qu'elles avaient au cours de la période antérieure. Mais certains thèmes enseignés auparavant ont disparu des programmes[17]. Déjà perceptible dans l'oeuvre d'Ibn al-Bannā, ce phénomène va s'amplifier à partir du XIVe siècle.

En second lieu, on ne décèle, dans ces commentaires, qu'un seul apport nouveau, celui de l'introduction, plus ou moins significative, d'un symbolisme mathématique relativement élaboré. Ce symbolisme n'est pas nouveau puisqu'on le rencontre déjà dans les écrits d'al-Ḥaṣṣar et d'Ibn al-Yāsamīn. Mais, il semble avoir été marginalisé tout au long du XIIIe siècle et durant la première moitié du XIVe siècle (du moins, nous n'en trouvons aucune trace dans les écrits qui nous sont parvenus), avant de réapparaître à la fin du XIVe et au cours du XVe siècle. En troisième lieu, on voit apparaître, à côté des manuels classiques utilisant le style et les formulations habituelles des mathématiques, un nouveau genre d'écrits où les commentaires grammaticaux, littéraires ou philosophiques occupent autant de places que les développements strictement mathématiques. L'exemple le plus typique est *Bughyat aṭ-ṭullāb* [Le désir des étudiants] d'Ibn Ghāzī (m. 1513)[18].

Tout ce que nous avons dit là concerne essentiellement la science du calcul. En géométrie, la seule certitude que nous avons concerne l'utilisation constante des " Eléments " d'Euclide et son enseignement. D'autres écrits importants, comme ceux d'Ibn as-Samḥ (m. 1034) et d'Ibn Munᶜim, ne semblent pas avoir circulé. Cela pourrait être un autre indice de l'abaissement du niveau dans l'enseignement de cette discipline. Mais notre information est encore trop lacunaire pour nous autoriser à être catégorique. Quoi qu'il en soit, on remarque que les seuls écrits qui sont évoqués par les mathématiciens postérieurs au XIVe siècle sont des écrits mineurs. Il s'agit des poèmes d'Ibn Liyūn (m. 1346) et d'Ibn ar-Raqqām (m. 1315)[19].

Dans le domaine de l'algèbre, différentes sources nous informent que les ouvrages d'Ibn Badr (XIIe s.), d'al-Qurashī, d'Ibn al-Yāsamīn et d'Ibn al-Bannā continuaient à être enseignés dans les villes du Maghreb, comme Fez, Tlemcen, Sebta et Tunis. Mais, au niveau des publications, seul le poème d'Ibn al-

17. Comme l'extraction de la racine cubique approchée d'un nombre ou le calcul de nouveaux couples de nombres amiables.

18. M. Souissi, *Bughyat aṭ-ṭullāb fī sharḥ Munyat al-ḥussāb li Ibn Ghāzī al-Miknāsī al-Fāsī* [Désir des étudiants sur le commentaire du " Voeu des calculateurs " d'Ibn Ghāzī al-Miknāsī al-Fāsī], Alep, Institut d'Histoire des sciences arabes, 1983.

19. M.L. Al-Khaṭṭābī, " Risālatān fī ᶜilm al-misāḥa li Ibn ar-Raqqām wa Ibn al-Bannā " [Deux épîtres d'Ibn ar-Raqqām et Ibn al-Bannā sur la science du mesurage], *Revue Daᶜwat al-ḥaqq*, n° 256 (Rabat, 1986), 39-47 ; M.L. Al-Khaṭṭābī, " Sharḥ al-Iksīr fī ᶜilm at-taksīr li Abī ᶜAbd Allāh Ibn al-Qāḍī " [Commentaire de l'élixir sur la science du mesurage d'Abū ᶜAbd Allāh Ibn al-Qāḍī], *Daᶜwat al-ḥaqq*, n° 258 (Rabat, 1987), 77-87.

Yāsamīn semble avoir eu la faveur des commentateurs[20].

LES MATHÉMATIQUES AU MAGHREB À L'ÉPOQUE OTTOMANE

Compte tenu de la situation de la recherche (encore modeste) sur les différentes activités mathématiques et astronomiques au Maghreb entre le milieu du XVIe et la fin du XIXe siècle, il n'est pas possible de présenter, à travers la production qui nous est parvenue, une analyse détaillée et une évaluation précise du contenu de toutes les disciplines ayant fortement utilisé les mathématiques, comme nous avions tenté de le faire, il y a une dizaine d'années déjà, pour l'Algèbre[21].

Nous nous contenterons donc de faire quelques remarques basées, essentiellement, sur des éléments biobibliographiques avec, parfois, le recours à des textes manuscrits ou lithographiés.

Les mathématiques d'expression arabe

Le nombre des auteurs connus ayant pratiqué les mathématiques ou l'astronomie, après le XVe siècle, et qui avaient un lien avec l'Ifriqya et le Maghreb Central (soit parce qu'ils y ont vécu et produit soit parce qu'ils en sont originaires), est estimé à une cinquantaine environ. En mathématique, leur production ou leur enseignement ont concerné la géométrie métrique, la science du calcul (incluant un chapitre d'algèbre), la construction des carrés magiques et la répartition des héritages. En astronomie, leurs préoccupations sont allées vers le calcul du temps (en particulier celui des prières quotidiennes et de la visibilité du croissant de lune), la détermination de la direction de la Mecque et, surtout, la description, la conception et l'utilisation d'instruments astronomiques.

En nous basant sur les sources que nous avons pu consulter jusqu'à ce jour, nous pouvons dire qu'au niveau du contenu, cette production s'inscrit dans le prolongement de celle du XVe siècle mais en accentuant les aspects qui distinguait déjà cette dernière de celle des XIIIe-XIVe siècles. Au niveau de la forme, on y distingue essentiellement trois types d'écrits bien différenciés mais qui expriment en fait une même matière scientifique et reflètent un même niveau. Le premier, et peut-être le plus ancien au Maghreb, est la versification. Elle utilise le mètre *rajaz,* appelé " l'âne des poètes ", parce qu'il est construit à partir de la répétition, dans chaque hémistiche du vers, d'un même rythme fonda-

20. J. Shawqī, *Manẓūmāt Ibn al-Yāsamīn fī a^cmāl al-jabr wa l-ḥisāb* [Les poèmes d'Ibn al-Yāsamīn sur les procédés de l'Algèbre et du Calcul], Koweit, Mu'assasat al-Kuwayt li t-taqaddum al-^cilmī, 1988.

21. A. Djebbar, " Quelques aspects de l'algèbre dans la tradition mathématique arabe de l'Occident musulman " dans *Actes du 1er Colloque Maghrébin d'Alger sur l'Histoire des Mathématiques Arabes, 1-3 Décembre 1986,* Alger, Maison du Livre, 1988, 99-123.

mental, *Mustafᶜilun* [pam, pam, papam]. Ce qui permettait aux auteurs qui n'avaient pas d'aptitude poétique spéciale, de reproduire leur discours mathématique sans trop de contrainte et avec un maximum de fidélité à son contenu technique.

Les poèmes les plus connus qui ont été produits au XVIe siècle sont, pour la région qui nous concerne ici, ceux d'al-Wansharīsī (m. 1549) et d'al-Akhḍarī (m. 1576), tous deux originaires du Maghreb Central.

Le premier de ces deux auteurs a mis en poème le fameux *Talkhīṣ aᶜmāl al-ḥisāb* d'Ibn al-Bannā en suivant en cela son professeur de Meknès Ibn Ghāzī dont la *Munyat al-ḥussāb* [Le voeu des calculateurs] est la plus célèbre versification du Talkhīṣ, à la fois pour sa forme, pour la valeur de son contenu et, surtout, pour le commentaire qui l'accompagne et qui est intitulé *Bughyat aṭ-ṭullāb* [Le désir des étudiants][22].

Quant à al-Akhḍarī, il a publié *ad-Durra al-bayḍa'* [La perle blanche] sur le calcul et la science des héritages, ainsi que le *Sirāj fi ᶜilm al-falak* [La lampe sur l'Astronomie]. Le contenu de ces deux poèmes confirme ce qui a été déjà dit sur le niveau scientifique des écrits du XVIe siècle. Pour prendre l'exemple de la *Durra al-bayḍā'*, on constate que son contenu mathématique est lui-même un abrégé de ce qui a été déjà résumé dans le *Talkhīṣ* d'Ibn al-Bannā, avec la disparition d'un certain nombre de notions théoriques et d'algorithmes. Ce qui est tout à fait compréhensible dans la mesure où le but d'al-Akhḍarī, comme celui de nombreux enseignants après lui, est de donner les outils mathématiques indispensables à la résolution des problèmes d'héritage. Cela apparaît d'ailleurs clairement dans la composition du poème puisque 400 vers y sont consacrés aux différents aspects des partages successoraux et seulement 130 vers pour l'exposé des instruments mathématiques[23].

Un des signes, révélateurs du changement qualitatif dans l'activité mathématique arabe après le XVe siècle (au niveau de tout l'Empire cette fois), est précisément la grande diffusion des deux poèmes d'al-Akhḍarī. En effet, en plus du nombre important de copies qui sont aujourd'hui conservées dans les bibliothèques du Maghreb, d'Egypte et de Turquie, il y a le nombre de commentaires dont elles ont bénéficié à partir de la fin du XVIe siècle. On trouve d'abord les commentaires des propres élèves d'al-Akhḍarī, comme al-Masbaḥ (m. 1572) pour la *Durra* et Ibn Muslim pour le *Sirāj*[24]. Pour les XVIIe-XVIIIe siècles, on peut citer Saḥnūn al-Wansharīsī, Muḥammad Ibn Ibrāhīm et Ibn al-

22. Nous n'avons encore trouvé aucune copie de ce poème et nous ne savons pas s'il a été utilisé ultérieurement.

23. A. Al-Akhḍarī, *ad-Durra al-bayḍā'* [La perle blanche], Le Caire, lithographie, 1891.

24. A. Saᶜdallah, *Tārīkh al-Jazā'ir ath-thaqāfī min al-qarn al ᶜāshir ilā al-qarn ar-rābi ᶜ ᶜashar al-hijrī* [Histoire culturelle de l'Algérie, du dixième au quatorzième siècle de l'Hégire], Alger, al-Mu'assasa al-waṭaniyya li l-kitāb, 2e éd., 1985, vol. II, 418.

Khabbāz[25], sans parler d'un certain nombre d'anonymes dont les commentaires nous sont parvenus[26]. La carrière de ces deux poèmes s'est poursuivie au XIXe siècle avec autant de succès, à la fois grâce à de nouveaux commentaires, comme celui d'Ibn Mallūka (m. 1859)[27], et à l'avènement de la lithographie qui va leur assurer une diffusion encore plus large[28].

Cette tendance à la versification des mathématiques n'est pas nouvelle puisque, dès le XIIe siècle, Ibn al-Yāsamīn publiait son poème algébrique suivi par un second sur les nombres irrationnels[29]. Mais le phénomène va s'amplifier à partir du XVIe siècle et plus particulièrement dans le Maghreb Extrême où nous avons recensé une quinzaine de poèmes mathématiques ou astronomiques écrits durant la période qui nous intéresse ici. Le plus connu de ces poèmes est le *Dhayl urjūzat ajniḥat ar-righāb fī l-ḥisāb* [Appendice au poème " les ailes des gens avides du calcul "] d'ar-Rasmūkī (m. 1726) qui a tout simplement complété un poème, écrit au XVe siècle par as-Samlali, en se limitant aux thèmes habituels de la science du calcul et des partages successoraux[30]. Les autres poèmes, moins connus et moins diffusés que le précédent, traitent des thèmes suivants : le calcul, chez Aḥmad Ibn al-Qāḍī (m. 1630)[31], les carrés magiques chez al-Fishtālī (m. 1649) et al-Marghīthī (m. 1679)[32], les héritages chez al-ᶜUqaylī (m. 1666), les instruments astronomiques chez Y. al-Fāsī (m.

25. S. Al-Wansharīsī, *Mufīd al-muḥtāj fī sharḥ as-Sirāj* [L'utile au nécessiteux dans le commentaire du *Sirāj*], Le Caire, lithographie, 1896 ; M. Ibn Ibrāhīm, *Sharḥ ad-Durra al-bayḍā'* [Commentaire sur la *Durra*], Ms. Tunis B.N. 4804/2 ; A. Ibn Al-Khabbāz, *al-Ghurra al-makhfiyya fī sharḥ ad-Durra al-alfiyya* [la première lueur cachée sur le commentaire de la *Durra* du millénaire], Ms. Tunis B.N. 3966.

26. Anonyme, *Sharḥ ad-Durra al-bayḍā'* [Commentaire sur la *Durra*], Ms. Tunis B.N. 345 ; Anonyme, *Muwaḍḍiḥ al-maᶜānī wa l-ashyā' ᶜalā abyāt ad-Durra al-bayḍā'* [Le clarificateur des idées et des choses sur les vers de la *Durra*], Ms. Alger B.N. 2133/1.

27. Pour ce commentateur et pour d'autres de la même époque, *cf.* M. Souissi, " Tadrīs ar-riyyāḍiyyāt bi l ᶜarabiyya fī l-Maghrib al-ᶜarabī wa khāṣṣatan bi Tūnus fī l-qarn ath-thālith ᶜashar wa n-niṣf al-awwal min al-qarn ar-rābiᶜ ᶜashar li l-Hijra " [L'enseignement des mathématiques en arabe dans le Maghreb arabe et en particulier en Tunisie au treizième siècle et dans la première moitié du quatorzième siècle de l'Hégire] dans *Actes du IIIe Colloque Maghrébin sur l'Histoire des Mathématiques Arabes, Alger, 1-3 Décembre 1990,* Alger, Office des Publications Universitaires, 1998, partie arabe, 33.

28. Deux exemples illustrent ce fait : la présence de la *Durra* dans le programme d'enseignement des mathématiques de la Zaytūna de Tunis (M. Souissi, " Tadrīs ar-riyyāḍiyyāt... ", *op. cit.*, 32) et son utilisation dans le manuel de Aḥmad Bābir al-Arwānī, un mathématicien de Tombouctou (Ms. Tombouctou, Bibl. Aḥmad Bābā, n° 3027, p. 7).

29. T. Zemouli, *al-Aᶜmāl ar-riyyāḍiyya li Ibn al-Yāsamīn* [L'œuvre mathématique d'Ibn al-Yāsamīn], Magister d'Histoire des Mathématiques, Alger, École Normale Supérieure, 1993.

30. Ar-Rasmūkī, *Miftāḥ ar-righāb fī maᶜrifat al-farā'iḍ wa l-ḥisāb* [La clé des gens avides à connaître les héritages et le calcul], Ms. Rabat Ḥasaniyya 5327. Le poème est accompagné du commentaire d'ar-Rasmūkī.

31. A. Ibn Al-Qāḍī, *Rajz Talkhīṣ Ibn al-Bannā* [versiflcation du Talkhīṣ d'Ibn al-Bannā].

32. Al-Fishtālī, *Urjūza fī l-wafq*, Ms. Rabat, B.H. 6675 ; Al-Marghīthī, *Manẓūma fī l-awfāq*, Ms. Rabat B.H. 5384. (cités par M. Lamrabet, *Introduction à l'histoire des mathématiques maghrébines,* Rabat, Al-maᶜārif al-jadīda, 1994, 148-149).

1680), M. al-Āsafī (m. 1720) et ^{c}A. al-Fāsī (m. 1685)[33], l'Algèbre chez ^{c}Ali Ibn al-Qāḍī (XVIIe s.)[34].

La seconde forme d'écrits est celle des résumés ou abrégés. Ce sont les moins nombreux et leur circulation n'a pas été très grande ; on peut signaler, par exemple, *ar-Risāla al-mukhtaṣara fi rub^{c} al-muqanṭarāt* [L'épître abrégée sur le quart des cercles de latitude] d'at-Tājūrī (m. 1584) et le *Mukhtaṣar salk ad-dārayn fī ḥall an-nayyirayn d'ar-Ru^{c}aynī* [L'abrégé de la voie des deux maisons sur la résolution des deux astres brillants] (m. 1584).

Quant à la troisième forme d'écrits, la plus courante pour la période concernée, elle comprend des gloses et, surtout, des commentaires. Certains de ces commentaires s'inscrivent dans une tradition qu'avait initiée ou popularisée, à Marrakech, le mathématicien Ibn al-Bannā puisqu'il avait conçu son important ouvrage *Raf^{c} al-ḥijāb ^{c}an wujūh amāl al-ḥisāb* [Le lever du voile sur les formes de procédés du calcul], comme un commentaire de son *Talkhiṣ* [L'Abrégé][35]. Au XVIe siècle, c'est une démarche analogue que l'on retrouve chez al-Akhḍari avec son *Sharḥ as-Sirāj* [Commentaire du Siraj]. Il fera d'ailleurs la même chose avec son livre de Logique, *as-Sullam* [L'échelle].

Mais, d'une manière générale, les commentaires portent sur des ouvrages d'autres auteurs. Pour les mathématiques, on peut citer, à titre d'exemples, les commentaires de Maqdīsh (XVIIIe s.), d'al-Majjājī (m. 1867) et du Shaykh Ṭfayyash (m. 1914) sur le *Kashf al-asrār ^{c}an ^{c}ilm ḥurūf al-ghubār* [Dévoilement des secrets de la science des chiffres de poussière][36]. En astronomie, il y a le *Sharḥ ^{c}alā qaṣīdat al-mujayyab* [Commentaire sur le poème du " cadrant " sinus] d'Ibn Ḥamādūsh (m. 1743) et les gloses d'al-Awmī (m. 1789) sur le poème astronomique d'Abū Miqrā^{c}.

A côté de ces trois formes particulières d'écrits scientifiques que sont les poèmes, les résumés et les commentaires, on trouve bien sûr des manuels ou des opuscules de facture classique. Comme on peut le constater au niveau des titres puis en parcourant le contenu de certains d'entre eux, cette dernière caté-

33. Y. Al-Fāsī, *Rajz fī l-asṭurlāb* [Poème sur l'astrolabe], Ms. Alexandrie, Ḥisāb 50 ; ^{c}A. Al-Fāsī, *Naẓm Risālat Ibn aṣ-Ṣaffār fī l-asṭurlāb* [Versification de l'épître d'Ibn aṣ-Ṣaffār sur l'astrolabe] ; M. Al-Āsafī, *Rajz fī manāzil al-qamar* [Poème sur les mansions lunaires], Ms. Rabat B.G. D1683 (cité par D. Lamrabet, *Introduction...*, *op. cit.*, 155).

34. ^{c}A. Ibn Al-Qāḍī, *as-Sirāj al-mubīn ilā urjūzat Ibn al-Yāsamīn* [La lampe éclairante sur le poème d'Ibn al-Yāsamīn], Ms. Rabat B.G. D 2133 (cité par D. Lamrabet, *Introduction...*, *op. cit.*, 147).

35. M. Aballagh, *Raf^{c} al-ḥijāb d'Ibn al-Bannā*, Thèse Doctorat, Paris, Université de Paris I-Pantheon Sorbonne, 1988 (Edition critique, traduction française et analyse mathématique).

36. Maqdīsh, *Ighāthat al-istibṣār ^{c}alā Kashf al-asrār ^{c}an ^{c}ilm ḥurūf al-ghubār* [L'aide à la compréhension pour le dévoilement des secrets de la science des chiffres de poussière], Ms. Tunis, B.N. 8154. Sur al-Majjājī, voir D. Lamrabet, *Introduction...*, *op. cit.*, 162. Sur Ṭfayyash, voir A. Djebbar, " Ba^{c}ḍ al-^{c}anāṣir ḥawla an-nashāṭāt ar-riyyāḍiyya fi l-Maghrib al-kabīr mā bayna al-qarnayn at-tāsi^{c} wa t-tāsi^{c} ^{c}ashar " [Quelques éléments sur les activités mathématiques au Maghreb entre le IXe et le XIXe siècle] (Séminaire National de Ghardaïa (Algérie), Avril 1993), dans *Tārikh ar-riyyāḍiyyāt al-^{c}arabiyya* (Alger, Association Algérienne d'Histoire des Mathématiques (éd.), 1996), 1-38.

gorie d'écrits, quantitativement la plus importante, ne traite pas de sujets originaux. En fait, et suivant en cela la tendance générale observée déjà à travers les trois autres catégories, les thèmes qui y sont traités se limitent aux aspects pratiques et calculables.

En mathématique, cela correspondait à l'apprentissage des quatre opérations arithmétiques (appliquées aux entiers, aux fractions et à certains nombres irrationnels) et des méthodes de résolution de problèmes à une inconnue (la règle de trois, la méthode de fausse position et la résolution algébrique). En astronomie, la formation reposait sur ce qui était suffisant comme bases géométriques et trigonométriques pour permettre la compréhension des principes régissant le calcul du temps ainsi que la conception ou l'utilisation des instruments les plus utilisés à cette époque : le quart de cercle, le quart de sinus, le cadran solaire et bien sûr les différents types d'astrolabes (locaux ou universels).

Quant au niveau de tous les écrits mathématiques et astronomiques arabes connus qui ont été produits entre le XVIe et le XIXe siècle, il se situe, sans exception, en deçà de celui des publications du XVe siècle tout en s'inscrivant dans la même orientation apparue à la fin du XIVe siècle et dont une des caractéristiques est l'exposé des techniques et des résultats sans aucune justification théorique. Mais, il faut préciser que, comme pour d'autres phénomènes déjà évoqués, celui du rétrécissement du domaine scientifique et de l'abaissement de son niveau n'est pas une spécificité du Maghreb ottoman. C'est absolument la même situation qui prévaut dans le Maghreb Extrême où, comme nous l'avons déjà signalé, l'on trouve les mêmes formes d'écrits scientifiques, les mêmes sujets et un niveau équivalent.

Traditions et expressions scientifiques non arabes au Maghreb

Dans tout ce qui précède, il ne s'est agi que des mathématiques et de l'astronomie, écrites et enseignées en langue arabe. Comme notre étude concerne une région qui a toujours pratiqué la langue amazigh[37], qui a également connu, ici ou là, l'usage du grec et du latin (au moins jusqu'à l'avènement de l'Islam) et qui, à partir du XVIe siècle, s'est enrichi du turc, il est normal de s'interroger sur le rôle de ces différentes langues dans la préservation d'un héritage mathématique ancien, ou dans la pratique scientifique elle-même, au cours de la période ottomane.

Au moyen âge, la production intellectuelle ou religieuse exprimée en amazigh a été écrite en utilisant l'alphabet arabe complété par des signes ajoutés à certaines lettres dans le but d'exprimer des phonèmes absents de la langue arabe[38]. Cette utilisation de l'amazigh a eu lieu très tôt si l'on en croit le témoignage du géographe andalou al-Bakrī (m. 1094) qui évoque une version

37. C'est le nom donné par les Maghrébins à la langue berbère.

38. Ibn Khaldūn est l'un de ceux qui ont eu à utiliser ce type de transcription pour respecter la prononciation des noms berbères. *Cf.* Ibn Khaldūn, *Kitāb al-^{c}ibar, op. cit.*, vol. I, 54-56.

du Coran exprimée dans cette langue sous la dynastie des Barghwāṭa (milieu du VIIIᵉ-milieu du XIᵉ siècle)[39]. Elle s'est poursuivie après cette date, en particulier avec l'avènement de la dynastie almohade dont les fondateurs, Ibn Tūmart (m. 1130) et ses successeurs immédiats, pratiquaient l'amazigh et l'écrivaient parfois tout en soutenant et en finançant une politique culturelle arabe[40]. Après eux, les dynasties berbères locales adopteront la même attitude consistant à parler l'amazigh et à pratiquer l'Arabe dans le domaine administratif, culturel et scientifique. C'est l'une des raisons (mais pas la seule) qui pourrait expliquer pourquoi l'utilisation de l'amazigh restera cantonnée, presque exclusivement, aux textes religieux ou poétiques[41].

C'est cette situation que l'on observe durant la phase ottomane du Maghreb, comme l'illustrent les écrits en amazigh d'al-Awzāl (m. 1748) dont les plus connus sont le *Baḥr ad-dumūᶜ* [L'océan de larmes][42] et *al-Ḥawḍ* [L'étang] une adaptation versifiée du *Kitāb al-Mukhtaṣar* [Le livre abrégé], le célèbre ouvrage de fiqh de Khalīl (m. 1374)[43].

Quant aux pratiques de calcul de la période gréco-latine du Maghreb, antérieure à l'avènement de l'Islam, il n'en a subsisté, semble-t-il, qu'un système de numération composé de 27 signes. Ce système, appelé " chiffres *rūmī* " ou " chiffres du *zimām* " ou " chiffres de *Fès* ", a continué à être utilisé dans certaines administrations au moins jusqu'au XVIIIᵉ siècle, comme le confirment les écrits d'al-ᶜUqaylī (m. 1665-66)[44], de ᶜAbd al-Qādir al-Fāsī (m. 1680) et de son commentateur Aḥmad Sakrīj (m. 1780 ?)[45]. Mais son utilisation semble s'être limitée au Maghreb Extrême, sans que l'on sache quelles ont été les véritables raisons de ce phénomène.

La langue turque dans les activités scientifiques des Maghrébins

Nous savons encore moins de choses, à l'heure actuelle, sur le rôle de la langue turque dans les activités scientifiques au Maghreb, entre le XVIᵉ et le XIXᵉ

39. Ch.-A. Julien, *Histoire de l'Afrique du Nord,* Paris, Payot, 1969, 39 ; A. Laroui, *L'histoire du Maghreb, op. cit.*, 104.

40. Selon Ibn al-Qaṭṭān, qui l'affirme dans son *Naẓm al-jumān*, Ibn Tūmart aurait écrit son *Kitāb at-tawḥīd* d'abord en amazigh. *Cf.* R. Bourouiba, *Ibn Tūmart*, Alger, Société Algérienne d'Édition et de Diffusion, 1974, 103.

41. Quelques rares textes scientifiques en amazigh sont signalés par des bibliographes, mais nous n'avons pas eu la possibilité de les consulter. *Cf.* A. Amahan, " Notes bibliographiques sur les manuscrits en langue tamazight écrits en caractères arabes ", dans A.Ch. Binebine (éd.), *Le manuscrit arabe et la codicologie,* Rabat, Université Mohamed V, 1994, 99-104, (Publications de la Faculté des Lettres et Sciences Humaines, Série Colloques et Séminaires, n° 33).

42. B.H. Stricker, *L'océan des pleurs, poème berbère de Muḥammad al-Awzāli*, Leiden, Brill, 1960.

43. A. Rahmani, *al-Ḥawḍ fī l-fiqh al-māliki bi l-lisān al-amāzighī* [L'étang sur le Droit malékite en langue amazigh], Casablanca, Binlid, 1977.

44. Al-ᶜUqaylī, *Ṣuwwar al-qalam ar-rūmī* [Formes de l'écriture de Fès], Ms. Rabat Hasaniyya 12032/11e, ff. 295b-302a.

45. Y. Güergour, " Les différents systèmes de numération au Maghreb à l'époque ottomane : l'exemple des chiffres rūmī " dans ce volume.

siècle. Son statut n'était évidemment pas comparable à celui de l'amazigh. En effet, cette dernière était une langue parlée et peu écrite qui avait, depuis le XIVe siècle, déserté les sphères du pouvoir, alors que le turc était la langue du pouvoir central et de ses institutions et qu'elle bénéficiait d'un double phénomène de diffusion : d'abord sur le plan géographique en transgressant les limites de l'Asie Mineure, à la faveur de l'extension de l'Empire ottoman et, surtout, sur le plan culturel et scientifique avec la redynamisation des activités mathématiques et astronomiques dans le centre de l'Empire.

Comme le montrent les derniers travaux biobibliographiques réalisés en Turquie[46], ce renouveau scientifique a connu deux phases bien distinctes au niveau de son contenu. Dans sa première phase, la langue turque a accompagné la langue arabe comme outil d'enseignement et de publication, comme l'avait fait le persan, en Asie Centrale à partir du XIe siècle. Ainsi, en plus des écrits scientifiques nécessaires à la vie quotidienne, comme les calendriers et certaines tables astronomiques, les spécialistes publient, en turc, des manuels, des commentaires sur des ouvrages écrits en arabe, des opuscules et des traités sur tous les sujets astronomiques encore au programme : Science du temps, instruments et tables astronomiques, etc.

Les *madrasa*, par leurs conceptions anciennes qui les liaient à leurs aînées de l'époque seljouqide, et par leur nombre (environ 200 dans la seule Anatolie) ont été, jusqu'au XVe siècle, un facteur non négligeable de consolidation et de pérennisation de l'enseignement des sciences en Arabe. Leur rôle dans la circulation de la production scientifique entre les différents foyers intellectuels de l'Empire n'est pas connu, même lorsque cette circulation est attestée, déjà au XIVe siècle, entre le Maghreb et l'Egypte[47]. Il est alors intéressant de savoir quel a été leur rôle dans la diffusion des ouvrages écrits en turc. L'étude plus poussée de l'histoire de ces institutions pourrait également nous éclairer sur la circulation des étudiants et des enseignants et sur les influences réciproques entre les foyers scientifiques du centre de l'Empire et ceux de province[48].

En attendant de disposer de nouvelles informations, nous allons nous contenter de quelques remarques. Il paraît d'abord raisonnable de penser que l'avènement du turc dans les activités scientifiques ne pouvait être ignoré par les communautés scientifiques et par les dirigeants du Maghreb, qu'ils aient été

46. E. Ihsanoğlu (éd.), *Osmanli astronomi literatürü tarihi*, [History of Astronomy Literature During the Ottoman Period], Istanbul, IRCICA, 1997.

47. A. Djebbar, " Transmission et échanges scientifiques en Méditerranée au temps des Croisades : l'exemple des mathématiques ". A paraître dans les *Actes du Colloque International sur l'Occident et le Proche-Orient au temps des Croisades : traductions et contacts scientifiques entre 1000 et 1300, Louvain-la-Neuve, 24-25 Mars 1997*(à paraître dans les Actes du colloque) ; M. Aballagh, " Introduction à l'étude de l'influence d'Ibn al-Bannā sur les mathématiques en Egypte à l'époque ottomane " dans ce volume.

48. E. Ihsanoglu, " Ottoman Science in the Classical Period and Early Contacts with European Science and Technology ", in *Transfer of Modern Sciences & Technology to the Muslim World,* Istanbul, 1992, 2-11.

ou non intégrés à l'Empire, et que cela a peut-être suscité de l'intérêt et des vocations.

C'est ce que l'on constate avec l'exemple de ^{c}Ali Ibn Ḥamza, un mathématicien originaire de la ville d'Alger. Son nom a souvent été cité, à tort, au sujet de l'élaboration du concept de logarithme. Mais son intérêt réside dans son profil de scientifique de la périphérie de l'Empire. Il a en effet rédigé, en turc, un ouvrage mathématique de très bonne facture pour l'époque, même si ses thèmes sont ceux que l'on rencontre habituellement dans l'un ou l'autre des ouvrages arabes antérieurs au XVIe siècle et que l'ordre de leur présentation rappelle certains ouvrages maghrébins, et plus particulièrement le *Talqīḥ al-afkār* d'Ibn al-Yāsamīn. En effet, après les définitions d'usage, Ibn Ḥamza consacre un chapitre aux quatre opérations arithmétiques sur les entiers, un second aux mêmes opérations appliquées aux fractions puis aux nombres irrationnels, un troisième aux procédés de résolution des problèmes à une inconnue (règle de trois, méthode de fausse position et procédés algébriques), un quatrième aux aires des figures planes et un dernier chapitre à un ensemble de problèmes d'application[49].

Il est probable que d'autres maghrébins aient eu une formation scientifique en turc et qu'ils aient exercé leurs métiers au Maghreb, comme Ibn Ḥamza semble l'avoir fait après son retour d'Istanbul. Mais nous ne pouvons pas encore le confirmer. Quoi qu'il en soit, il ne semble pas que l'on est en présence d'un phénomène s'inscrivant dans le prolongement de celui du Centre de l'Empire. La première raison est probablement liée à la place du turc dans l'enseignement de base au Maghreb dans la mesure où ni le pouvoir central ni les pouvoirs locaux ne semblent avoir pris d'initiatives déterminantes dans ce sens. La seconde raison est à chercher dans la situation encore dominante de la science d'expression arabe dans les foyers les plus dynamiques de tout l'espace ottoman et plus particulièrement à Istanbul dont la vitalité scientifique a attiré des mathématiciens et des astronomes de différentes régions de l'Empire qui s'y sont perfectionnés ou qui y ont enseigné et publié en arabe. Pour l'Ifriqya, on peut citer Muḥammad Maghūsh (m. 1540), considéré comme un spécialiste des sciences rationnelles, ainsi que les deux astronomes Muḥammad al-Bārī (XVIIe s.) et Muḥammad al-Makkī (m. 1817)[50].

49. S. Zeki, *Āthār bāqiyya* [Vestiges restantes]. Cité par Q.H. Ṭūqān, *Turāth al-^{c}Arab al-^{c}ilmī fi r-riyāḍiyyāt wa l-falak* [Le patrimoine scientifique arabe en Mathématique et en Astronomie], Beyrouth, Dār ash-shurūq, 1963, 470-471.

50. Pour al-Bārī, *cf.* E. Ihsanoğlu (éd.), *Osmanli astronomi literatürü tarihi, op. cit.*, vol. I, 338. Pour Maghūsh, *cf.* A.B. At-Tunbuktī, *Nayl a-ibtihāj bi taṭrīz ad-Dībāj* [L'obtention de l'allégresse par l'ornementation du Dibaj], Beyrouth, Dār al-kutub al-^{c}ilmiyya, s.d., 336. Pour al-Makkī, *cf.* A. Al-^{c}Azzāwī, *Tārīkh ^{c}ilm al-falak fi l-^{c}Iraq wa ^{c}alaqatuhū bi l-aqṭār al-islāmiyya wa l-^{c}arabiyya fi l-^{c}uhūd at-tāliyya li ayyām al-^{c}Abbāsiyyīn (1258-1917)* [Histoire de l'Astronomie en Irak et ses liens avec les régions islamiques et arabes aux époques postérieures à celle des Abbassides (1258-1917)], Bagdad, Maṭba^{c}at al-majma^{c} al-^{c}ilmī al-^{c}irāqī, 1958, 331.

Il reste, malgré tout cela, une interrogation à propos de l'écart significatif que l'on observe entre le niveau scientifique des villes les plus actives du Maghreb et celui des villes comme Istanbul ou Konya, même si l'innovation au sens des IXe-XIIe siècles, est absente dans l'une et l'autre des deux traditions. En effet, lorsque la comparaison est possible, c'est-à-dire lorsqu'on se limite à la production d'expression arabe, on constate que, si l'on peut mettre sur le même plan les commentaires de la *Durra al-bayḍā'* d'al-Akhḍarī, réalisés par des Maghrébins, et ceux de la *Khulāṣat al-ḥisāb* [La conclusion du calcul] d'al-ᶜĀmilī (m. 1622) réalisés par des Turcs, cela n'est plus possible lorsqu'on veut comparer, par exemple, l'un des ouvrages astronomiques de Mīrim Çelebī (m. 1525) ou de Taqiy ad-Dīn Ibn Maᶜrūf (m. 1585) au *Sirāj* de leur contemporain al-Akhḍarī[51]. Pour le XVIIe siècle, l'écart est encore plus flagrant si on en juge par les écrits d'Ibn Ḥamādūsh, un intellectuel du Maghreb Central représentatif de son milieu et de son époque[52].

L'appropriation de la science européenne et ses prolongements au Maghreb

Le second aspect de la dynamisation des activités scientifiques dans l'Empire ottoman est en totale rupture avec le premier. Il englobe toutes les initiatives qui vont naître et se développer autour de ce qu'on pourrait appeler l'appropriation sélective de la science et de la technologie européenne par les institutions et les milieux scientifiques de l'Empire[53]. Cette appropriation, puissamment encouragée et soutenue par le pouvoir central, a été bien étudiée à travers ses aspects liés à la Turquie proprement dite.

On sait qu'elle a concerné, dans un premier temps, la cartographie, les technologies militaires et la médecine. Puis, les traductions se sont étendues à des ouvrages d'astronomie et de mathématiques comme ceux de Cassini, Clairaut et de Lalande, pour ne citer que les plus connus. Elle s'est également poursuivie, à la fin du XVIIIe siècle, par la mise en place d'un enseignement scientifique nouveau dans de nouvelles institutions : l'Ecole de Mathématique (1733), l'Ecole impériale des ingénieurs navals (1773) et l'Ecole impériale des ingénieurs militaires (1795)[54].

On sait également que ce phénomène a été relayé par certains milieux intellectuels de Turquie et d'Egypte. Quant à son prolongement au Maghreb, il n'a

51. Sur ces deux grands astronomes du XVIe siècle, *cf.* E. Ihsanoğlu (éd.), *Osmanli astronomi literatürü tarihi, op. cit.*, vol. I, 90-101 (pour Çelebi), 199-217 (pour Ibn Maᶜrūf).

52. A. Saᶜdallah, " Risala fi l-kura al-falakiyya li Ibn Hamadush " [Epître d'Ibn Hamadush sur la sphère] dans *IIIe Colloque Maghrébin sur l'Histoire des Mathématiques Arabes*, *op. cit.*, partie arabe, 23-30.

53. A. Al-ᶜAzzāwī, *Tārīkh ᶜilm al-falak fi l-ᶜIraq...*, *op. cit.*, 281-285.

54. E. Ihsanoğlu, " Introduction of Western Science to the ottoman World : A Case Study of Modem Astronomy (1660-1860) ", in E. Ihsanoğlu (éd.), *Transfer of Modern Science..., op. cit.*, 67-120 ; F. Günergun, " Introduction of the Metric System to the Ottoman State ", in *Transfer in Modern Science..., op. cit.*, 297-316.

pas encore été étudié pour lui-même. Mais, pour le domaine qui nous intéresse ici, et au vu des informations dont nous disposons sur les profils des hommes de science et sur le contenu de leur production, il semble que, jusqu'au milieu du XIXe siècle, la situation n'a pas sensiblement évolué. En effet, en dehors de quelques initiatives (comme celles d'Ibn Ḥamādūsh qui méritent une étude particulière)[55], c'est la tradition scientifique du XVe siècle qui se perpétue mais avec, comme nous l'avons déjà souligné, un abaissement général du niveau et un rétrécissement significatif du champ des connaissances. Cette situation est relativement plus accentuée au Maghreb Central comme le révèle le contenu de la production mathématique et astronomique de cette région, comparée à celle de l'Ifriqya et du Maghreb Extrême[56].

C'est d'ailleurs dans cette dernière région qui, à l'exception de quelques péripéties, était toujours restée hors de la mouvance politique ottomane, que l'on observe, dans le domaine des activités scientifiques, un début de changement qualitatif allant dans le sens des initiatives ottomanes[57]. Un processus d'appropriation de la science et de la technologie européenne se met en place et se développe, encouragé par les initiatives et le mécénat du sultan Muhammad IV lui-même (1859-1873) : traduction en arabe d'ouvrages scientifiques français (*L'application de l'Algèbre à la Géométrie* de G. Monge et J.-N. Hachette, *La bibliographie astronomique* de J. Lalande, *La Géométrie* de Legendre), création d'une Ecole de Géométres à Fès, envoi d'étudiants en Europe[58].

Cela dit, quel qu'il soit l'impact de ce phénomène de transfert scientifique dans l'une ou l'autre région du Maghreb au cours de la seconde moitié du XIXe siècle, il n'a jamais eu l'importance qu'on lui a connue en Turquie et en Egypte. Parmi les indices allant dans le sens de cette remarque, il y a la prudence et souvent le conservatisme qui ont caractérisé le programme d'enseignement des deux grandes mosquées-universités, celle d'al-Qarawiyyīn à

55. A. Sacdallah, *Ibn Ḥamādūsh al-jazā'irī, ḥayātuhū wa āthāruhū* [L'algérien Ibn Ḥamādūsh, sa vie et son œuvre], Alger, Office des Presses Universitaires, 1982, 53-56. Parmi les écrits de ce médecin-polygraphe, l'auteur signale les titres suivants : (1) *Ta'līf fī l-qaws alladhī ya'khudhu bihī an-Naṣārā* [Ecrit sur l'arc qu'utilisent les chrétiens], (2) *Ta'līf can ar-rukhāma aḍ-ḍilliyya bi l-ḥisāb istakharajahū min kutub an-Naṣārā* [Écrit sur le cadran-tangente à l'aide du calcul, extrait des livres des Chrétiens]. Ces titres révèlent une circulation, directe ou indirecte, de certains écrits scientifiques européens dans la périphérie de l'Empire ottoman. Il serait intéressant de voir quel a pu être le lien entre cette circulation et l'encouragement prodigué, par le pouvoir et certains milieux intellectuels ottomans, à la diffusion de la science européenne.

56. A. Sacdallah, *Tārīkh al-Jazā'ir ath-thaqāfī, op. cit.*, vol. II, 416-429.

57. M. Al-Manūnī, " Namādhij min tafattuḥ Maghrib al-qarn 19 calā mucṭayāt nahḍat Ūrubbā wa sh-Sharq al-islāmī " [Quelques exemples de l'ouverture du Maroc du XIXe siècle sur les données de la renaissance de l'Europe et de l'Orient musulman] dans *Actes du Colloque sur la Réforme et société marocaine au XIXe siècle, Rabat, Faculté des Lettres, 20-23 Avril 1983*, 193-203.

58. M. Al-Manūnī, *Maẓāhir yaqẓat al-Maghrib al-ḥadīth* [Aspects de la renaissance du Maroc moderne], Beyrouth, Dār al-Gharb al-islāmī, 1985.

Fès[59] et celle d'az-Zaytūna à Tunis. Pour prendre l'exemple de cette dernière institution, il faut rappeler que qu'une réforme a bien été élaborée et officialisée par une loi de 1876 mais son application a été dénaturée, en particulier en rendant optionnelle l'étude des matières scientifiques nouvelles[60].

Il y a aussi, pour le Maghreb Extrême, la résistance qui s'est manifestée avec l'avènement de la lithographie[61]. Cette technologie, qui était une sorte de transition vers l'imprimerie, a été perçue par certains milieux comme une *bid*ᶜ*a* [innovation blâmable]. Pourtant, à y regarder de plus près, on constate que ce procédé, révolutionnaire pour l'époque, n'a pas porté un projet complètement novateur dans la mesure où il n'a pas accompagné, et encore moins favorisé, des initiatives modernistes dans le domaine des sciences. On constate, au contraire, qu'elle est restée, presque exclusivement, au service de la culture traditionnelle. En effet, parmi les centaines d'ouvrages qui ont bénéficié de la lithographie, il y eut d'abord les grands classiques de l'orthodoxie malékite maghébine, c'est-à-dire le *Muwaṭṭa'* [Le livre facilité] de l'Imām Mālik (m. 795), la *Risāla* [L'épître] d'Ibn Abī Zayd al-Qayrawānī (m. 996) et le *Shifā'* [La guérison] du Qāḍī ᶜIyyāḍ (m. 1149).

Il faudra attendre 1876 pour que le premier livre de mathématique bénéficie d'une impression lithographiée. Il s'agit de la rédaction des *Eléments* d'Euclide publiée, six siècles auparavant en Asie Centrale, par Naṣīr ad-Dīn aṭ-Ṭūsī (m. 1274)[62].

59. M. Al-Manūnī, " Faṣla taṣifu ad-dirāsa bi l-Qarawiyyīn ayyām al-Manṣūr as-Saᶜdī " [Note décrivant l'enseignement à la Qarawiyyine à l'époque du saᶜdide al-Manṣūr], *Revue al-Bahth al-ᶜilmī*, n° 7 (1966), 241-266.

60. M. Souissi, *Tadrīs ar-riyyāḍiyyāt..., op. cit.*, partie arabe, 34-35.

61. M. Bencheneb, E. Levi-Provençal, *Essai de répertoire chronologique des éditions de Fès*, Alger, 1922 ; H.P.-J. Renaud, " L'enseignement des sciences exactes et l'édition d'ouvrages scientifiques au Maroc avant l'occupation européenne ", *Archeion*, vol. 13 (1931).

62. A.-Ch. Binebine, *Histoire des bibliothèques au Maroc*, Rabat, Publications de la Faculté des Lettres, 1992, 199 (Série Thèses et mémoires, n° 17).

Les différents systèmes de numérotation au Maghreb à l'Époque ottomane : l'exemple des chiffres *rūmī*

Youcef Guergour

Introduction

A notre connaissance, cinq systèmes de numération étaient employés au Maghreb et ce dès le début du XIIe siècle. Il s'agit des systèmes utilisant :

1. Les chiffres de poussière (ḥurūf al-ghubār)[1], appelés de nos jours chiffres arabes, comportant, comme on le sait, neuf symboles représentant les nombres de un à neuf, en plus d'un dixième symbole représentant le zéro. La plupart des traités parus en Occident Musulman, à partir du XIIe siècle, font usage de ces symboles.

2. Les chiffres qu'on appelle indiens (ḥurūf al-Hind), qui étaient utilisés essentiellement en Orient, et qui comportent eux aussi neuf symboles pour les chiffres de un à neuf et un point pour le zéro. On les trouve dans les manuels de calcul, d'Algèbre et de Géométrie.

3. La numération alphabétique (ḥurūf al-jummal). Ce système utilise l'alphabet arabe qui compte 28 lettres. A chaque lettre est associée soit une des neuf unités, soit une des neuf dizaines soit une des neuf centaines, la vingt-huitième lettre étant associée à mille, une valeur qui ne dépend pas de sa position dans la suite de lettres représentant un nombre. En passant d'Orient en Occident, cette numérotation a connu quelques permutations entre certaines lettres de l'alphabet[2].

1. Ibn al-Yāsamīn, " Kitāb Talqīh al-afkār fī l-ᶜamal bi rushūm al-ghubār " [La fécondation des esprits par le calcul à l'aide des chiffres de poussières], dans T. Zemouli, *Les écrits mathématiques d'Ibn al-Yāsamīn (m. 601/1204),* Magister, Alger, École Normale Supérieure, 1993, 9.

2. Les lettres suivantes sont les mêmes en Orient et en Occident : a, b, g, d, h, w, z, ḥ, ṭ, y, k, l, m, n (1, 2, 3, 4, 5, 6, 7, 8, 9, 10, 20, 30, 40, 50). A partir de là, la numérotation orientale qui est ainsi : s, a, f, ṣ, q, r, sh, t, th, kh, dh, ḍ, ẓ, gh (60, 70, 80, 90, 100, 200, 300, 400, 500, 600, 700, 800, 900, 1000), est modifiée en Occident comme suit : ṣ= 60, ḍ= 90, s = 300, ẓ = 800, gh = 900, sh = 100.

4. Le calcul digital ou calcul de la main (ḥisāb al-yad) : la numération et les opérations arithmétiques utilisent les doigts de la main qui servent à désigner les unités, les dizaines, les centaines, les milliers et les dizaines de milliers[3]. Ce calcul a été exposé dans certains manuels maghrébins. On le trouve évoqué, dès le XIIe siècle, par Ibn Munᶜim (m. 626/1228) dans son *Traité Fiqh al-ḥisāb* (La science du calcul). Après avoir défini les chiffres *rūmī*, il ajoute : “ Si tu veux travailler avec les chiffres *rūmī*, il faut en premier lieu connaître le calcul digital et apprendre l'indice et les noms des nombres. Et si tu veux travailler avec les chiffres *ghubār*, tu n'as pas besoin d'apprendre les indices et les noms des nombres ”[4].

5. Les chiffres *rūmī* : il semble que ces chiffres ont subi plusieurs transformations avant de devenir les *chiffres de Fès* (al-qalam al-fāsī), également appelés les chiffres des registres (ḥurūf z-zimām)[5]. Ces chiffres comportent 27 symboles et leur système de numération est un système décimal non positionnel. On trouve ces chiffres dans les documents notariaux anciens (legs, estimation d'indemnités) ainsi que dans les transactions commerciales. Le premier maghrébin connu qui leur a consacré une étude épître séparée est Ibn al-Bannā (m. 721/1321).

Le but de cette courte étude est de présenter une description et une analyse de cette épître qui est intitulée *al-Iqtiḍāb min al-ᶜamal bi r-rūmī fī l-ḥisāb.*

PRÉSENTATION ET ANALYSE DE L'ÉPÎTRE D'IBN AL-BANNĀ SUR LES CHIFFRES *RŪMĪ*[6]

Du titre même de l'épître, on peut déduire que ces chiffres étaient utilisés couramment dans les transactions commerciales et dans d'autres transactions de la cité islamique du Maghreb. En effet, le terme *Iqtiḍāb* signifie “ extrait ” (au sens d'abréviation, de raccourcissement) de ce qui est abondamment pratiqué. D'ailleurs, cette utilisation courante des chiffres *rūmī* ne date pas de l'époque d'Ibn al-Bannā, si on en croit le mathématicien du XIIe siècle al-Ḥaṣṣar (XIIe s.) qui évoque dans son *Kitāb al-Kāmil* [Le livre complet] l'utilisation de ces chiffres par les services administratifs de l'époque[7]. Cette situation n'a pas dû sensiblement changer après Ibn al-Bannā puisque ces chiffres sont évoqués par ses commentateurs du XIVe siècle et que, après le XVe siècle, des auteurs plus modestes vont leur consacrer des petites épîtres.

3. Un certain nombre d'écrits a été consacré à ce système de numération et à son utilisation dans les calculs. A titre d'exemple, on peut citer le poème d'Ibn al-Maghribī, intitulée *Urjūza fī ḥisāb al-yad* (Ms. Gotha 1495).

4. Ibn Munᶜim, *Fiqh al-ḥisāb*, Ms. Rabat, B.G., n° 416 Q, p. 220.

5. M. Al-Manouni, *Nashāṭ ad-dirāsāt ar-riyyāḍiyYa fī Maghrib al ᶜaṣr al-waṣīṭ ar-rābiᶜ al-Manahil,* Rabat, n° 33 (1985), 77-115.

6. Ibn Al-Bannā, *al-Iqtiḍāb min al-ᶜamal bi r-rūmī fī l-ḥisāb*, Ms. Rabat, B.B., n° 416 K, 425-432.

7. Al-Ḥaṣṣār, *al-Kitāb al-Kāmil*, Marrakech, B. Ibn Yūsuf, Ms. n° 313, ff. 9a-11b.

L'épître d'Ibn al-Bannā comporte cinq chapitres. Le premier traite des noms des positions (ordres), des chiffres et de leurs symboles qui sont représentés ainsi dans la copie qui nous servi pour cette étude :

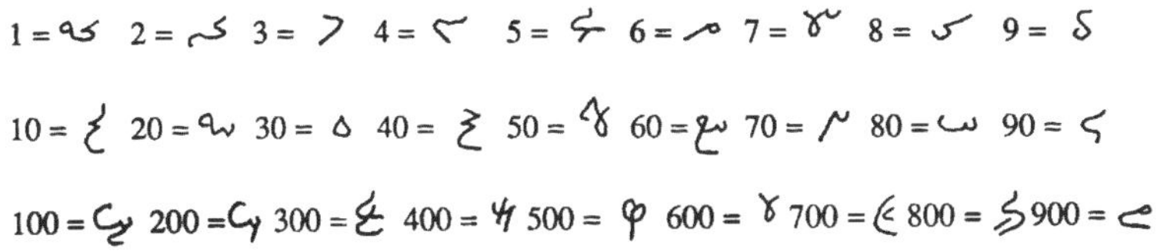
1 = 2 = 3 = 4 = 5 = 6 = 7 = 8 = 9 =
10 = 20 = 30 = 40 = 50 = 60 = 70 = 80 = 90 =
100 = 200 = 300 = 400 = 500 = 600 = 700 = 800 = 900 =

A partir de la position des milliers, on retrouve ces mêmes symboles mais soulignés d'un ou de plusieurs traits : la position des milliers utilisant un seul trait, celle millions deux traits, et ainsi de suite. On a, par exemple :

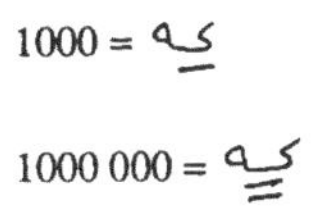

Ibn al-Bannā expose ensuite les symboles des fractions : chacune d'elle est représentée par le chiffre souligné dont elle est dérivée, au-dessus duquel on représente le chiffre correspondant aux parties de la fraction. Par exemple, six septièmes s'écrit ainsi (en ce qui concerne certains quantièmes, Ibn al-Bannā signale que les spécialistes de ce domaine ont convenu d'utiliser les écritures suivantes) :

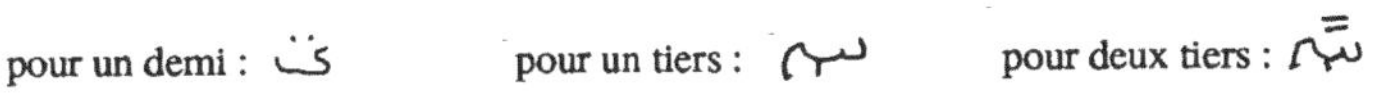
pour un demi :
pour un tiers :
pour deux tiers :

Remarquons que les quantièmes dont les dénominateurs dépassent 10 ne sont pas utilisés. Lorsque les spécialistes dans ce domaine rencontrent quelque chose de ce genre, ils le décomposent à l'aide de l'addition en quantièmes de dénominateurs inférieurs à 10.

Ibn al-Bannā classe les fractions[8] en deux types seulement, celles qui sont dites " rapportées " (*mudafa*), c'est-à-dire les fractions de fractions, et celles qui sont dites " différentes " (*mukhtalifa*), c'est-à-dire les fractions simples ou les sommes de fractions simples et rapportées[9].

Le second chapitre de l'épître est consacré à l'addition. Dans son livre *al-Talkhīṣ*, Ibn al-Bannā définit l'addition comme l'opération " consistant à réunir

8. Dans la tradition mathématique maghrébine, les fractions étaient classées selon les types suivants : fractions élémentaires (simple, continue, ou discontinue), fractions hétérogènes (continue ou discontinue). *Cf.* A. Djebbar, *Enseignement et recherche mathématiques dans le Maghreb des XIIIe-XIVe siècles*, Paris, Université de Paris-Sud, 1980 (Publications Mathématiques d'Orsay, n° 81-02), 46-47.

9. A. Djebbar, " Le traitement des fractions dans la tradition mathématique arabe du Maghreb ", dans P. Benoît, K. Chemla, J. Ritter (éds), *Actes du Colloque International sur l'Histoire des fractions, Paris, 30-31 Janvier 1987*, Bâle, Boston, Berlin, Birkhauser Verlag, 1992, 223-245 (" Histoire de fractions, fractions d'histoire ").

les nombres les uns aux autres afin de pouvoir les exprimer par un nom unique "[10]. Ici, l'addition est définie, implicitement, comme l'opération qui consiste à regrouper les noms représentant les deux nombres à sommer. L'auteur fait remarquer que les spécialistes dans ce domaine commencent par le plus grand nombre, puisque c'est le plus important à leurs yeux, vu son ordre de grandeur.

Toujours dans le *Talkhīṣ*, Ibn al-Bannā expose l'addition pour les entiers, les fractions et les nombres quadratiques en distinguant cinq types d'addition[11], tandis que dans l'*Iqtiḍāb*, il se contente d'étudier deux cas : celui où la somme porte sur des entiers et celui où elle porte sur des entiers et des fractions.

Si la somme ne fait intervenir que des entiers, il pose les deux nombres sur deux lignes superposées et effectue l'opération comme nous le faisons maintenant en commençant par les unités. Il illustre la méthode en faisant la somme de trois cent quatre-vingt-cinq et huit cent quatre-vingt-quatre. Le résultat, soit mille deux cent soixante-dix-neuf, est placé soit au-dessus des deux nombres, soit en-dessous.

Si la somme porte sur des entiers et des fractions, les entiers sont sommés ensemble, comme dans le premier cas, et les fractions ensemble selon la règle en usage à cette époque.

Dans le troisième chapitre, l'auteur traite de la soustraction, en la définissant comme étant le reste du résultat de l'opération consistant à déduire un nombre d'un autre. Ibn al-Bannā précise que, pour faciliter cette opération le calculateur est amené, quelques fois, à ajouter ou retrancher la même quantité aux deux termes de la soustraction[12], par exemple :

$$14-8 = 16-10 = 6 \qquad (a-b = (a+c)-(b+c))$$

L'auteur ajoute que les calculateurs avec ce système de numération procèdent de la manière suivante : à chaque étape de l'opération, ils font appel à un nombre qui, ajouté à celui qu'on veut soustraire, donne le nombre dont on soustrait. Le nombre à ajouter sera alors le résultat cherché. Voici, comme ils procèdent pour soustraire quatre-vingt-cinq de quatre-vingt-seize : en commençant par les unités, on cherche le nombre qui, ajouté à cinq donne six. C'est ici un que l'on fixe dans la position des unités. Puis on passe à la position

10. M. Souissi, *Ibn al-Bannā' al-Murrāhushī, Talkhīṣ a^{c}māl al-ḥisāb* [Ibn al-Bannā' de Marrakech, L'abrégé des opérations du Calcul] (édition, traduction française et commentaires), Tunis, Publications de l'Université de Tunis, 1969, 44.

11. Addition de nombres entre lesquels il n'existe pas de relation connue, addition de la suite naturelle des nombres, de leurs carrés et de leurs cubes, addition de la suite des nombres impairs, de leurs carrés et de leurs cubes, addition de la suite des nombres pairs, de leurs carrés et de leurs cubes.

12. Dans le *Talkhīṣ,* Ibn al-Bannā définit la soustraction comme suit : " La soustraction consiste à rechercher le reste obtenu en ôtant l'un des deux nombres de l'autre. Elle est de deux sortes : l'opération où l'on soustrait le nombre le plus petit du plus grand une seule fois, celle où l'on soustrait le plus petit du plus grand, plus d'une fois jusqu'à épuisement de ce dernier ou qu'il subsiste un reste inférieur au plus petit ".

des dizaines : si on ajoute dix à quatre-vingts on aura quatre-vingt-dix. On fixe dix dans la position des dizaines. Il reste alors onze. Le reste de l'opération est alors onze.

Le quatrième chapitre traite de la multiplication[13]. Avant d'aborder l'algorithme de la multiplication proprement dit, Ibn al-Bannā suggère au calculateur d'apprendre soigneusement la règle suivante qui donne l'ordre de grandeur du produit : " Si les deux facteurs sont formés seulement d'unités, le produit est de l'ordre des dizaines ; s'ils sont dans la position des dizaines, alors le produit sera dans la position des centaines ; s'ils sont dans la position des centaines alors le résultat sera dans la position des dizaines de milliers ; si l'un des deux est dans la position des dizaines et l'autre dans la position des centaines alors le produit sera dans la position des milliers, et ainsi de suite. Si, dans les deux nombres du produit, ou dans l'un d'eux, il y a des répétitions, on fait l'opération sans les répétitions, puis on remet les répétitions ".

Il faut remarquer que l'auteur ne donne pas, dans cette épître, la définition de la multiplication, comme il le fait dans le *Talkhīṣ*[14]. Il aborde directement l'algorithme, à travers des exemples. Voici comment il procède : pour multiplier six mille trois cent vingt-cinq par deux cent seize mille deux cent quatre-vingts. On pose les deux nombres à multiplier sur deux lignes. Ensuite, on effectue la multiplication par 5, puis par 2, puis par 3 et finalement par 6, de la manière suivante : 5x80 = 40 dizaines ; 5x2 = 10 centaines = mille ; 5x6 = 30 mille, etc.

Puis Ibn al-Bannā expose le procédé de multiplication des fractions, selon une méthode analogue à celle utilisée à l'aide les chiffres Ghubār.

Le chapitre cinq traite de la division[15]. Ibn al-Bannā commence par faire la remarque suivante : " Sache que la division utilisant " les chiffres " *rumi* exige beaucoup d'intuition pour arriver au quotient, qui est le nombre dont la multiplication par le diviseur redonne le dividende ". Cette remarque de l'auteur laisse à penser que la division était considérée par les utilisateurs comme une opération délicate. Dans cette même remarque, on trouve une définition de la division différente, dans sa formulation de celles qu'il donne dans le *Talkhīṣ*[16].

13. Dans son livre *Ḥaṭṭ an-niqāb*, Ibn Qunfudh dit que " la multiplication avec ces chiffres est très délicate ". Voir Y. Guergour, *Les écrits mathématiques d'Ibn Qunfudh, op. cit.*, 39.

14. Il y est dit que " la multiplication consiste à répéter l'un des deux nombres autant de fois qu'il y a d'unités dans le second ".

15. Ibn Qunfudh dit, au sujet de la division : " Tu n'as pas besoin de l'apprendre, car si tu as compris la multiplication, tu devrais la comprendre puisque ses opérations se feront avec la multiplication ". Voir Y. Guergour, *Les écrits mathématiques d'Ibn Qunfudh, op. cit.*, 39.

16. La définition du *Talkhīṣ* est comme suit : " La division consiste à décomposer le dividende en autant de parties égales qu'il y a d'unités dans le diviseur. On entend (également) par division le rapport de l'un des deux nombres à l'autre ".

LA PLACE DES CHIFFRES *RŪMĪ* AU MAGHREB

Dans un article daté de 1933, l'historien des sciences G.S. Colin a donné des arguments en faveur de l'origine grecque des chiffres *rūmī* en précisant, qu'ils ont été transmis au Maghreb extrême à partir de l'Asie mineure puis de l'Espagne musulmane[17]. Ce qui renforce cette hypothèse c'est l'existence de ces chiffres chez des mathématiciens arabes antérieurs au XIVᵉ siècle, comme al-Uqlidisi qui les évoque dans son traité *al-Fuṣūl fī l-ḥisāb al-hindī*[18]. Mais nous n'avons, pour le moment, aucune information concernant l'utilisation de ces chiffres dans les activités quotidiennes des cités d'Orient.

Pour l'occident musulman, les informations que nous possédons, aujourd'hui, nous permettent de dire que ces chiffres ont circulé et ont été utilisés couramment au Maghreb, et plus particulièrement dans le Maghreb Extrême.

Parmi les mathématiciens maghrébins qui ont utilisé ou évoqué ces chiffres, il y a d'abord Abū Bakr al-Ḥaṣṣār qui en donne, dans son traité *al-Kāmil fī ᶜilm al-ᶜadad*, la forme et la manière de les utiliser. A propos de l'écriture des grands nombres, il suggère, pour éviter des erreurs de lecture, de substituer aux petits traits soulignant les milliers par la lettre de la numération alphabétique correspondant au nombre de traits.

Après lui, ou peut-être à la même époque, des mathématiciens d'origine andalouse ont également évoqué ces chiffres. Il s'agit, en particulier, d'Ibn Munᶜim dans son livre *Fiqh al-Ḥisāb* et d'Ibn Bundūd (XIIᵉ s.) dans son grand traité de calcul qui ne nous est pas parvenu mais dont nous connaissons certains aspects grâce au témoignage du mathématicien Ibn Zakariyā' al-Gharnāṭī (808/1406).

A partir du XIVᵉ siècle, les chiffres continuent à être traités dans trois types d'écrits mathématiques au moins : des commentaires du *Talkhīṣ* d'Ibn al-Bannā, des épîtres entièrement consacrées à ces chiffres et des *urjūza* (poèmes mathématiques).

Pour la catégorie des commentaires, il y a le *Ḥaṭṭ an niqāb ᶜan wujūh aᶜmāl al-ḥisāb* [L'abaissement de la voilette sur les opérations du calcul] d'Ibn al-Qunfudh (m. 810/1407)[19] dans lequel l'auteur s'intéresse aux symboles de ces chiffres ainsi qu'aux opérations élémentaires réalisées sur les entiers et les fractions : addition, soustraction, multiplication, division. De très brèves indications sur les propriétés de ces chiffres apparaissent aussi dans le traité *Ḥaṭṭ an niqāb baᶜda rafᶜ al-hijab ᶜan wujūh aᶜmāl al-ḥisāb* [L'abaissement de la

17. G.S. Colin, " De l'origine Grecque des chiffres de Fès et de nos chiffres arabes ", *Journal Asiatique* (avril-juin 1993), 193-215.

18. A.S. Saidan, *Al-Fūsūl fī l-ḥisāb al-hindī*, Alep, I.H.A.S., 1985, 95.

19. Y. Guergour, *Les écrits mathématiques d'Ibn Qunfudh al-Qasanṭīnī* (810/1407), Magister en Histoire des Mathématiques, Alger, E.N.S, 1990.

voilette après le soulèvement du voile sur les opérations du calcul] d'Ibn Zalcaryya' al-Gharnati[20]. L'un des dernier commentaires qui évoque ces chiffres est *Bughyat aṭ-ṭullāb fī sharḥ Munyat al-ḥussāb* [Désir des étudiants sur le commentaire du voeu des calculateurs] d'Ibn Ghāzī al-Miknāsī (m. 919/1513)[21].

Les épîtres sur l'utilisation des chiffres *rūmī* qui ont été écrites après celle d'Ibn al-Bannā s'inscrivent dans la même tradition et sont les témoins de la survivance de la pratique du calcul à l'aide de ces chiffres. La plus connue d'entre elles est celle d'al-ᶜUqaylī (m. 1076/1665-66), intitulée *Ṣuwwar al-qalam ar-rūmī wa ᶜamalihi wa mabda' ṣurati āḥādihī* [Les formes des chiffres *rūmī*, leur utilisation et le principe de la forme de ses unités][22].

Quant aux poèmes, nous ne connaissons, pour l'instant, que celui de ᶜAbd al-Qādir al-Fāsī (m. 1091/1680), intitulé *Guide de celui qui veut apprendre et de celui qui a oublié, à propos de la forme des chiffres de Fès*. Ce poème sera d'ailleurs commenté par un Aḥmad Sakrīj dans un ouvrage intitulé *Irshād al-mutaᶜallim an-nāsī fī ṣifat ashkāl al-qalam al-fāsī* [Guide de l'apprenant oublieux sur la forme des figures des chiffres de Fès][23].

CONCLUSION

En guise de conclusion, il nous semble utile de faire les remarques suivantes :

1. Dans le système de numération *rūmī*, on ne trouve aucune forme pour représenter le zéro.
2. Il est connu qu'à l'époque d'Ibn al-Bannā plusieurs algorithmes servaient à calculer le produit de deux nombres. Or, dans son épître sur les chiffres *rūmī*, Ibn al-Bannā ne mentionne qu'un seul algorithme.
3. Le silence de l'auteur laisse à penser que cette numération ne servait pas à des opérations faisant intervenir des nombres irrationnels quadratiques ou biquadratiques. D'ailleurs aucun des auteurs qui ont évoqué les chiffres *rūmī* ne font allusion à des champs d'application autres que celui des entiers et celui des fractions.
4. Il semble que ces chiffres n'ont pas circulé hors d'al-Andalus et du Maghreb extrême, même après le XIVᵉ siècle. En effet, les seuls auteurs connus qui les ont évoqués à cette époque, sont Ibn Qunfudh et Ibn Zakariyya'. Le

20. Ms. Tunis, B.N., n° 561, p. 9.

21. M. Souissi, *Bughyat aṭ-ṭullāb fī sharḥ Munyat al-ḥussāb li Ibn Ghāzī al-Miknāsī al-Fāsī* [Désir des étudiants sur le commentaire du voeu des calculateurs d'Ibn Ghāzī al-Miknāsī al-Fāsī], Alep, Institut d'Histoire des sciences arabes, 1983.

22. Ms. Rabat, B.H., n° 12302/11ᵉ, ff. 295b-302a.

23. Ahmad Sakrīj, *Irshād al-mutaᶜallim an-nāsī fī ṣifat ashkāl al-qalam al-fāsī* [Guide de l'apprenant oublieux sur la forme des figures des chiffres de Fès], Fez, s.d., édition lithographiée. Traduction française de Charles Pellat, Alger, 1917.

premier est andalou. Quant au second, bien qu'il soit originaire de Constantine, il a vécu presque vingt ans à Fès et on sait qu'il y a rédigé son commentaire *Ḥaṭṭ an-niqāb*.

5. Les écrits traitant des chiffres *rūmī* ne semblent pas avoir été destinés à l'enseignement académique. Ce sont plutôt des compléments pour cet enseignement ou des manuels s'adressant à des utilisateurs déjà intégrés dans les activités quotidiennes de la cité maghrébine, telles que les transactions commerciales, les répartitions des héritages, les partages de bénéfice, etc.

Introduction à l'étude de l'influence d'Ibn al-Bannā sur les mathématiques en Egypte à l'époque ottomane

Mohamed Aballagh

Introduction

La transmission des mathématiques grecques et orientales de l'Orient musulman vers l'Occident musulman a suscité plus d'intérêt que l'étude du phénomène inverse, c'est-à-dire la présence, en Orient, des écrits mathématiques rédigés en Occident musulman.

Pourtant, avec les travaux de J.P. Hogendijk sur al-Mu'taman Ibn Hūd (m. 1085)[1] et ceux de A. Djebbar sur la tradition mathématique arabe de l'Occident musulman, un début d'intérêt pour cette question commence à voir le jour.

C'est ainsi qu'il est connu actuellement, que grâce à Maïmonide (m. 1205), le grand livre *al-Istikmāl* (la perfection) d'al-Mu'taman était enseigné puis commenté en Egypte[2]. De même, pour le poème algébrique du mathématicien maghrébin Ibn al-Yasāmīn qui a fait l'objet de commentaires orientaux, parmi lesquels on peut citer celui qui a été réalisé par Ibn al-Hā'im al-Miṣrī (m. 1412)[3].

Il est certain aussi que les travaux d'autres mathématiciens de l'Occident musulman des XIe et XIIe siècles étaient connus en Orient et plus particulièrement en Egypte. Nous avons, par exemple, trouvé dans un ouvrage encyclopédique rédigé par un égyptien du XVe siècle l'évocation du livre *al-Kāmil* du

1. Voir la liste de ses travaux sur ce mathématicien dans A. Djebbar, " La rédaction de l'Istikmāl d'al-Mu'taman (XIe s.) par Ibn Sartāq, un mathématicien des XIIIe-XIVe siècles ", *Historia Mathematica,* 24 (1997), 192.

2. A. Djebbar, " Deux mathématiciens peu connus de l'Espagne du XIe siècle : al-Mu'taman et Ibn Sayyid ", dans *Vestigia Mathematica : Studies in Medieval and Early Modern Mathematics in Honour of H.L.L. Busard, M. Folkerts & J.P. Hogendijk,* Amsterdam, Atlanta, Rodopi, 1993, 79-91.

3. T. Zemouli, *Les travaux mathématiques d'Ibn al-Yāsamīn,* Magister en Histoire des Mathématiques, Alger, École Normale Supérieure, 1993, 12. L'auteur y évoque aussi d'autres noms de commentateurs orientaux : Sibṭ al-Māradīnī (m. 1501) et al-cIrāqī (m. 1423).

mathématicien andalou Ibn as-Samḥ (m. 1034) ainsi que celui du mathématicien maghrébin al-Ḥaṣṣār qui a vécu au XIIe siècle[4].

Ces informations montrent bien qu'il y avait, avant le XIVe siècle, entre l'Ouest et l'Est du monde arabo-musulman, des liens solides que des études plus approfondies pourraient expliciter. Par contre, nous sommes mieux informés sur la transmission des livres mathématiques du Maghreb vers l'Egypte au XIVe et au XVe siècle.

Comme nous allons le voir dans le paragraphe suivant, l'une des raisons principales de ce fait est le succès des travaux mathématiques d'Ibn al-Bannā et son influence sur les mathématiciens de l'Occident musulman et de l'Égypte.

L'INFLUENCE DES ÉCRITS MATHÉMATIQUES D'IBN AL-BANNĀ SUR LES MATHÉMATIQUES DES XIVe-XVe SIÈCLES

Nous avons recensé dans une précédente étude, en collaboration avec A. Djebbar, seize écrits mathématiques d'Ibn al-Bannā[5]. Parmi eux, quatre livres ont été le plus utilisés. Il s'agit de *Arbaᶜ maqālāt fī l-ᶜadad* [Les quatre épîtres sur le nombre][6], *al-Uṣūl wa l-muqaddimāt fī l-jabr wa l-muqābala* [Les fondements et les préliminaires en Algèbre][7], le *Talkhīṣ aᶜmāl al-Ḥisāb* [L'abrégé des opérations du Calcul][8] et le *Rafᶜ al-ḥijāb ᶜan wujūh aᶜmāl al-Ḥisāb* [Le lever du voile sur les différentes opérations du Calcul][9].

Mais, seuls les deux derniers ont été commentés. C'est ainsi que nous avons recensé, pour le *Talkhīṣ*, dix-sept grands commentaires, quatre versifications et un résumé[10], le *Rafᶜ al-ḥijāb* n'ayant, quant à lui, fait l'objet que d'un seul commentaire[11].

4. Al-Qalqashandī, *Ṣubḥ al-aᶜshā fī kitābat al-inshā'* [La lumière de l'aveugle dans l'art de la rédaction], Le Caire, Imprimerie impériale, vol. I, 478-480.

5. M. Aballagh, A. Djebbar, *Ḥayāt wa mu'allafāt Ibn al-Bannā* [La vie et l'oeuvre d'Ibn al-Bannā], Publications de la Faculté des Lettres de Rabat (à paraître).

6. A.S. Saidan, *Tārīkh ᶜilm al-ḥisāb ᶜinda l-ᶜArab, al-juz' ath-thālith : al-ḥisāb fī l-Andalus wa l-Maghrib* [Histoire de la Science du calcul chez les Arabes, Troisième partie : le Calcul en Andalousie et au Maghreb], Amman, Dār al-Furqān, 1984.

7. A. Djebbar, *Mathématiques et Mathématiciens du Maghreb médiéval (IXe-XVIe siècles) : Contribution à l'étude des activités scientifiques de l'Occident musulman,* Thèse de Doctorat, Université de Nantes-Université de Paris-Sud, 1990, vol. II (partie arabe).

8. M. Souissi, *Ibn al-Bannā' al-Murrāhushī, Talkhīṣ aᶜmāl al-ḥisāb* [Ibn al-Bannā' de Marrakech, L'abrégé des opérations du Calcul] (édition, traduction française et commentaires), Tunis, Publications de l'Université de Tunis, 1969.

9. M. Aballagh, *Rafᶜ al-ḥijāb d'Ibn al-Bannā*, Thèse Doctorat, Université de Paris I-Panthéon-Sorbonne. Édition critique, traduction française et analyse mathématique. Version arabe : Fès, Publication de la Faculté des Lettres, n° 5, 1994.

10. M. Aballagh, A. Djebbar, *Ḥayāt wa mu'allafāt Ibn al-Bannā, op. cit.*

11. Ibn Haydūr at-Tādlī, *Tuḥfat aṭ-ṭullāb fisharḥ mā ashkala min Rafᶜ al-ḥijāb* [Parure des étudiants sur l'explication de ce qui est problématique dans le *Rafᶜ al-ḥijāb*], Ms. Vatican, n° 1403.

Le *Talkhīṣ,* qui a été écrit au cours de la dernière décennie du XIIIe siècle[12], est donc le plus célèbre ouvrage mathématique d'Ibn al-Bannā. Il l'est également par rapport à toute la tradition mathématique de l'Occident musulman postérieure au XIIIe siècle.

Ce livre traite des quatre opérations arithmétiques classiques sur les entiers et les fractions (addition, soustraction, multiplication et division), des racines carrées, des quatre grandeurs proportionnelles, de la méthode de fausse position et de l'algèbre, c'est-à-dire une matière semblable à celle que nous trouvons dans d'autres manuels de calcul antérieurs au XIIIe siècle.

Mais, ce qui caractérise le *Talkhīṣ,* c'est la démarche qui y est suivie par Ibn al-Bannā : il ne donne pas d'exemples pour illustrer les règles et les procédés exposés et il se contente d'indiquer la méthode générale de résolution. Les questions traitées sont nombreuses mais le texte ne contient pas de répétition. La matière étudiée est variée et chaque proposition apporte une nouvelle notion et un procédé différent[13].

Ce caractère concis du *Talkhīṣ* et la politique d'enseignement de l'époque[14] ont incité les mathématiciens à préférer cette démarche dans le domaine de l'enseignement et dans celui de la rédaction d'ouvrages mathématiques.

Parmi les dix-sept grands commentaires du *Talkhīṣ,* certains ont été écrits par des mathématiciens maghrébins, quelques-uns par des andalous et un seul par un égyptien, qui n'est autre qu'Ibn al-Majdī (m. 1446). Quant aux résumés du *Talkhīṣ,* on ne connaît que celui du mathématicien égyptien Ibn al-Hā'im.

Pour mieux apprécier l'oeuvre d'Ibn al-Majdī, qui fera l'objet du troisième paragraphe de cet article, il faudrait bien sûr le comparer aux autres commentaires. Malheureusement, et jusqu'à ce jour, aucune étude globale concernant les commentaires du *Talkhīṣ* n'a été réalisée, en particulier parce que l'édition de ces textes n'a pas encore été achevée[15].

Nous allons donc rappeler, ici, les différentes méthodes utilisées par les commentateurs, ce qui sera un moyen pour les évaluer. On constate d'abord que, tandis que des commentateurs se contentent de donner des exemples illus-

12. M. Aballagh, A. Djebbar, *Ḥayāt wa mu'allafāt Ibn al-Bannā, op. cit.*

13. M. Aballagh, *La tradition du savoir dans Fès médiévale,* Rabat, 1992, 65 (Collection Autrement ; Série Mémoires, 13).

14. M. Kably, " Qaḍiyyat al-madāris al-marīniyya, mulāḥaḍāt wa ta'amulāt " [La question des Madrasas mérinides, remarques et réflexions], dans *Murāja^c^at ḥawla l-mujtama^c^ wa th-thaqāfa bi l-Maghrib al-wasīṭ* [Revue de la société et de la culture dans le Maroc médiéval], Casablanca, Éditions Toubkal, 1987.

15. Il faudrait, à ce propos, signaler l'édition critique, réalisée à l'ENS d'Alger sous la direction de A. Djebbar, de deux commentaires du *Talkhīṣ* d'Ibn al-Bannā, celui d'Ibn Qunfudh et celui d'al-^c^Uqbānī. *Cf.* Y. Guergour, *al-A^c^māl ar-riyyāḍiyya li Ibn Qunfudh al-Qasanṭīnī (m. 810/1407)* [Les écrits mathématiques d'Ibn Qunfudh al-Qasanṭīnī (m. 810/1407)], Magister d'Histoire des Mathématiques, Alger, École Normale Supérieure, 1990 ; A. Harbili, *L'enseignement des mathématiques à Tlemcen au XIVe siècle à travers le commentaire d'al-^c^Uqbānī (m. 811/1408),* Magister en Histoire des mathématiques, Alger, École Normale Supérieure, 1997.

trant les règles générales du *Talkhīṣ*, d'autres mathématiciens appuient leurs propos par des démonstrations justifiant ces propositions. Et, tandis que des mathématiciens négligent, ou parfois ne font que paraphraser, le *Rafᶜ al-ḥijāb*, ouvrage jugé difficile à l'époque, d'autres mathématiciens l'utilisent pour approfondir le contenu du *Talkhīṣ*. Parfois, on trouve même des critiques à l'encontre de certaines affirmations d'Ibn al-Bannā dans le *Talkhīṣ* et dans le *Rafᶜ al-ḥijāb*.

IBN AL-MAJDĪ ET LA TRADITION MATHÉMATIQUE MAGHRÉBINE DANS L'ÉGYPTE PRÉ-OTTOMANE

Le résumé du *Talkhīṣ* d'Ibn al-Hā'im, intitulé *al-Ḥāwī* [Le recueil] a été rédigé avant qu'Ibn al-Majdī n'entreprenne la rédaction de son *Ḥawī l-lubāb fī sharḥ aᶜmāl al-Ḥisāb* [Le recueil de la moelle des procèdés du calcul][16]. Puisque le *Ḥawī* d'Ibn al-Hā'im a fait l'objet, à son tour, de commentaires et de versification, on peut raisonnablement penser que c'est bien par le biais de ces deux mathématiciens que la tradition mathématique maghrébine sera connue dans l'Egypte ottomane.

Mais, comme nous ne disposons pas de copie du résumé écrit par Ibn al-Hā'im, nous allons, dans cette brève étude, nous contenter de signaler quelques caractéristiques du commentaire d'Ibn al-Majdī, qui nous est parvenu à travers plusieurs copies[17].

Dans son étude, réalisée en 1980, sur les mathématiques dans le Maghreb des XIIIᵉ-XIVᵉ siècles, A. Djebbar avait consacré plusieurs paragraphes au contenu du livre d'Ibn al-Majdī[18].

A partir de cette étude, on peut affirmer qu'Ibn al-Majdī est l'un des meilleurs commentateurs du *Talkhīṣ*. En effet, que ce soit dans le domaine de l'Algèbre, ou dans celui de la Théorie des nombres et du Calcul, on remarque que ce mathématicien ne se contente pas de bien commenter l'aspect novateur du *Rafᶜ al-ḥijāb* mais il étend les résultats de ce dernier livre à d'autres domaines mathématiques.

Par exemple, Ibn al-Bannā utilise les outils algébriques dans l'étude des suites arithmétiques, tandis qu'Ibn al-Majdī étend l'utilisation de l'instrument algébrique à l'étude des suites géométriques[19].

16. Ibn al-Majdī a écrit son commentaire après la mort d'Ibn al-Hā'im, comme le confirme le paragraphe dans lequel il se réfère au commentaire de ce dernier sur le poème algébrique d'Ibn al-Yāsamīn. *Cf.* Ibn al-Majdī, *Ḥāwī l-lubāb*, Ms. British Museum, n° 7469, (dernière partie non paginée).

17. M. Aballagh, A. Djebbar, *Ḥayāt wa mu'allafāt Ibn al-Bannā, op. cit.*

18. A. Djebbar, *Enseignement et Recherche mathématiques dans le Maghreb des XIIIᵉ-XIVᵉ siècles*, Paris, Université de Paris-Sud, 34 (Publications Mathématiques d'Orsay, n° 81-02).

19. *Op. cit.*, 34.

Il faut également signaler que l'importance donnée, par Ibn al-Bannā, aux problèmes de dénombrement trouvera un grand écho et un prolongement dans le commentaire d'Ibn al-Majdī[20].

En effet, ce dernier se distingue des autres commentateurs par son originalité dans l'utilisation des procédés et des démarches du nouveau chapitre des mathématiques qui est l'analyse combinatoire.

C'est ainsi qu'il dénombre les opérations élémentaires qui interviennent dans les produits sans translations et par semi-translation lorsque les deux nombres à multiplier sont identiques[21]. Dans les chapitres de l'addition, il donne à propos des arrangements des lettres de l'alphabet, un complément et une généralisation que nous ne trouvons pas chez Ibn al-Bannā[22].

Mais, l'importance du commentaire d'Ibn al-Majdī ne réside pas seulement dans ce que nous avons dit, mais aussi par le fait qu'il a été enrichi par des éléments de mathématiques orientales que nous n'avons pas repérés dans les écrits mathématiques de l'Occident musulman qui nous sont parvenus.

C'est ainsi qu'il est le seul, à notre connaissance, à rejeta catégoriquement toute utilisation de la Philosophie dans un texte mathématique[23].

A différents endroits de son commentaire, Ibn al-Majdī évoque des critiques adressées à certaines formulations d'Ibn al-Bannā dans son *Talkhīṣ*. Nous n'avons pas retrouvé ces critiques dans les commentaires connus de la tradition mathématique de l'Occident musulman que nous avons consultés. Ce qui nous autorise à pensa qu'Ibn al-Majdī n'est peut-être pas le seul mathématicien égyptien, ou oriental, à avoir commenté le *Talkhīṣ* d'Ibn al-Bannā.

Nous avons ainsi constaté que quelques critiques évoquées par Ibn al-Majdī sont au niveau de la terminologie employée par Ibn al-Bannā. Par exemple, ce dernier parle d'une méthode de produit appelée " méthode couchée ". Ibn al-Majdī fait remarquer que des commentateurs ont trouvé étrange l'emploi du terme " couché " dans un texte mathématique[24].

En consultant les commentaires réalisés en Occident musulman, particulièrement ceux d'Ibn Qunfudh, d'Ibn Haydūr et d'al-Maṣrātī, nous n'avons trouvé aucune critique concernant l'emploi de ce terme. Cela tendrait à prouver que son utilisation était familière dans le milieu mathématique du Maghreb, ce qui ne semble pas avoir été le cas, à la même époque, en Egypte.

20. *Op. cit.*, 72.

21. *Op. cit.*, 97.

22. *Op. cit.*

23. Voir sur la présence de la Philosophie dans quelques textes mathématiques au Maghreb : A. Djebbar, " Quelques remarques sur les rapports entre Philosophie et Mathématiques arabes " (Colloque de la Société Tunisienne de Philosophie, Hammamet, 1-2 Juin 1983), dans *Revue Tunisienne des Etudes Philosophiques*, n° 2 (Mars 1984), 3-21 ; M. Aballagh, *Rafᶜ al-ḥijāb d'Ibn al-Bannā, op. cit.*, 75-93.

24. Ibn al-Majdī, *Ḥāwī l-lubāb, op. cit.*, f. 39b.

Signalons enfin que, tandis qu'aucun auteur maghrébin n'évoque la résolution des équations de degré supérieur ou égal à trois (à l'exception d'une évocation rapide d'Ibn Khaldūn dans sa Muqaddima)[25], Ibn al-Majdī réserve l'annexe de son commentaire à la classification et à l'énumération des équations de degré supérieur à 2[26]. Ce qui montre qu'Ibn al-Majdī connaissait certains aspects de la tradition mathématique orientale dans ce domaine.

Tous les éléments que nous avons rapportés dans ce paragraphe montrent l'importance de ce commentaire d'Ibn al-Majdī qui peut être considéré comme une synthèse des mathématiques de l'Orient et de l'Occident musulman au XVe siècle.

CONCLUSION

Il est désormais possible de dire que ce travail réalisé au XVe siècle, c'est-à-dire un siècle avant que l'Egypte ne soit intégrée dans l'Empire ottoman, va faire des mathématiques de l'Occident musulman, et plus particulièrement de celles du Maghreb, l'une des trois composantes des mathématiques de cette longue histoire de l'Egypte, puisqu'il faut ajouter à la synthèse des deux traditions arabes d'Orient et d'Occident, celle de l'Europe qui va pénétra en Égypte grâce aux traductions de certains manuels qui vont être réalisées au XIXe siècle, dans le sillage de l'important phénomène de traduction, en turc, d'ouvrages de différentes disciplines scientifiques[27].

Les informations que nous possédons actuellement montrent, en effet, que cette synthèse à laquelle a contribué Ibn al-Majdī s'est perpétuée à travers ses étudiants qui vont poursuivre son enseignement en utilisant la matière de son livre ou en rédigeant de nouveaux commentaires, comme l'a peut-être fait son élève Ibn al-Warrāq[28].

Signalons enfin que le résumé du *Talkhīṣ* rédigé par Ibn al-Hā'im a connu, lui aussi, des prolongements, par exemple le poème et le grand commentaire qui lui ont été consacrés par Ibn aṣ-Ṣayrafī (m. 1500)[29].

25. A. Djebbar, *Enseignement et Recherche mathématiques dans le Maghreb des XIIIe-XIVe siècles, op. cit.,* 5.

26. *Op. cit.,* 107-112.

27. E. Ihsanoğlu, " Les Ottomans et la science européenne ", dans S. Onen, C. Proust (éds), *Les écoles savantes en Turquie, sciences, philosophie et arts au fil des siècles*, Istanbul, Éditions Isis, 1996, 165-171.

28. Ibn al-Majdī, *Ḥāwī l-lubāb, op. cit.*, f. 38a. Le nom de ce mathématicien est cité en marge. A cause de la similitude entre le titre de l'écrit d'Ibn al-Majdī et celui d'Ibn al-Hā'im, on ne peut pas, à partir de cette seule note marginale, trancher sur la question de savoir si Ibn al-Warrāq avait commenté le livre déjà très détaillé de l'un ou le résumé de l'autre. Mais, il nous semble plus probable qu'il ait commenté celui d'Ibn al-Hā'im.

29. M. al-Manūnī, " Nashāṭ ad-dirāsāt ar-riyāḍyya fī Maghrib al-ᶜaṣr al-waṣīṭ ar-rābiᶜ " [Activité des études mathématiques dans le Maroc de la quatrième période du moyen-âge], *Al-Manhil*, n° 33 (1985), 77-115.

THE DEVELOPMENT IN THE ATTITUDE OF THE OTTOMAN STATE TOWARDS SCIENCE AND EDUCATION AND THE ESTABLISHMENT OF THE ENGINEERING SCHOOLS (*MÜHENDISHANES*)

Mustafa KAÇAR

Until the mid-17th century, the Ottomans had political, economic and military superiority over both the Western and Eastern worlds[1]. It was during the Ottoman siege of the island of Crete (1645-1669) that European armies became a serious threat for the Ottomans for the first time. The Ottoman rulers realized that the difficulties they experienced in this conquest were mainly due to the inefficient military techniques and tactics they employed. Eager to benefit from military innovations, Ottoman rulers employed European and non-Muslim Ottoman experts on firearms and new methods of fortification[2].

Ottoman intellectuals such as Katip Çelebi (d. 1657) and Ibrahim Müteferrika (d. 1745) also put forward their opinions about the deficiencies in the Ottoman administration and military organization in the various memorandums (*layihas*) they prepared. These intellectuals identified the weaknesses in the state apparatus and warned the rulers about them. Ibrahim Müteferrika, a Unitarian and a convert to Islam, was the first to clearly express the need for change, and tried to draw the attention of the Ottoman ruling class to the military developments taking place in Europe. Müteferrika, the founder of the first Turkish language printing house in 1728[3], pointed out that reform was necessary both among the ruling class and within the society at large. He argued that the failure of the Ottoman armies against European powers was the result of

1. E. Ihsanoglu, " Modernisation Efforts in Science, Technology and Industry in the Ottoman Empire (18th and 19th Century), F. Günergun (ed.), *The Introduction of Modern Science and Technology to Turkey and Japan*, Shigehisa Kuriyama, Kyoto, International Research Center for Japanese Studies, 1998, 15-35.

2. L.F. Marsigli, *L'état Militaire de l'Empire Ottoman,* Part II, Amsterdam 1732, in facsimile, Graz Austria 1972, 33.

3. N. Berkes, *The Development of Secularism in Turkey*, Montreal, McGill University, 1964, 36-37.

rapid developments in European military technology and enlightened European administration[4]. In the Ottoman army, the introduction of similar techniques began with the organization of the Bombardiers.

Realizing that military defeats mainly resulted from outdated battle techniques, the Ottomans first attempted to organise the Bombardiers after the European model. Thus, the most noteworthy reforms in the Ottoman military organisation took place in the area of military technical training.

An Ottoman corps of bombardiers had existed since the 15th century, attached to the Kapikulu infantries. Until the beginning of the 18th century their organization included soldiers with technical expertise on the bombing of distant invisible targets using mortars. Bombardiers were divided into three sections : armories (*cebeci),* artillerymen (*topçu*) and fief holders (*timarli*) who were all administered by a Chief Bombardier. While the Chief Bombardier held an office in Istanbul, members of the corps of bombardiers were employed at border fortresses. Bombardiers of the time were not paid salaries but were granted fiefs in return for their service[5]. Comte Marsigli, an Italian nobleman who came to Istanbul as a war prisoner at the end of the 17th century, described the corps of bombardiers in his book on the Ottoman military organisation. He reported that the Turks had mortars of various types and magnitudes, but were unable to use them properly. According to Marsigli, Ottoman rulers considered the art of bombardiership very important and thus employed a Venician officer to train soldiers as bombardiers[6].

THE FOUNDATION OF THE *ULUFELI HUMBARACI OCAGI* (CORPS OF SALARIED BOMBARDIERS)

The foundation of the *Ulufeli Humbaraci Ocagi* was a first step in reforming the Ottoman army on the European model. This corps of salaried bombardiers was founded in 1735 under the supervision of General Claude Alexandre de Bonneval (1675-1747), a French general.

Comte de Bonneval, known to the Turks as Humbaraci Ahmed Pasha, was the 3rd son of a noble French Family. He owed his worldwide reputation to his distinguished military achievements. He has a special place in the Ottoman history due to his role in the adoption of European war techniques to the Ottoman army[7]. He took refuge in the Ottoman Empire in Sarajevo, embraced Islam in 1729, and took the name of Ahmed. During his stay in Bosnia he wrote two important letters : one to Damat Ibrahim Pasha of Nevsehir, then the Grand

4. I. Müteferrika, *Usûlü'l- Hikem fi Nizâmü'l-Ümem,* Darut-tibaat-i Amire, Konstantiniyye 1144/1731, p. 16b.

5. I. Hakki Uzunçarsili, *Osmanli Devleti Teskilatindan Kapikulu Ocaklari*, I, 2nd edition, Ankara 1984, 117-127.

6. L.F. Marsigli, *L'état Militaire de l'Empire Ottoman*, Part II, 33.

7. *Biographie Universelle*, vol. 5, Paris 1812, 135-136.

Vizier and the other to Villeneuve, the French Ambassador to the Porte. In the first letter he glorified the majesty of the Ottoman Sultan and the Grand Vizier, and expressed his wish to serve in the Ottoman army[8]. In his second letter, addressed to the Ambassador, he talked about the purpose of his voyage and stated that his aim was to cooperate with the Turks and regain the territories of Hungary by defeating the Germans. Probably his real goal was to carve out a kingdom in Hungary[9].

Bonneval was forced to live in Bosnia for one year. He was called to the palace in 1730. However, as soon as he got to Edirne (300 km from Istanbul), the *Patrona Halil* revolt broke out and he was ordered to wait in Gümülcine, a small town near Edirne. The revolt was quickly suppressed ; Sultan Ahmed III was dethroned and Mahmud I succeeded to the throne. Bonneval then sent a letter to the new Sultan reiterating his wish to serve in the Ottoman army. He put his personal seal at the end of the letter, as was the custom in Ottoman official correspondence. Having introduced himself in detail, Bonneval enumerated his military victories achieved in France, Italy and Austria. He also explained that his intention in taking refuge in the Ottoman State was to fight against the enemies of Ottoman State. He added that he had a lot of information about the enemy and held many maps of enemy lands that could be of use to the Ottomans[10].

Topal Osman Pasha, the new Grand Vizier, invited Bonneval to Istanbul for the second time on October 21, 1731. The aim was to benefit from his experience in the application of modern military techniques. Bonneval arrived in Istanbul in February 1732. He met the Grand Vizier as soon as he arrived and was appointed as Chief Bombardier. At that time it was unusual for a convert to Islam to hold this post[11].

At the request of the Grand Vizier, Bonneval prepared a detailed report on the military power of some of the European kingdoms, including France, Holland, England, and Spain. This report contained information about the military organizations of these kingdoms, as well as the various training methods and techniques employed in their armies. The most significant aspect of this report was that it contained information on the training given in French military schools, emphasizing the importance of engineers knowledgeable in mathematics employed within the French army. This report was influential both in Bonneval's subsequent activities and in the preparation of the regulation for the

8. Letter dated 28 June 1729, send from Sarajevo, Septime Gorseix, *Bonneval Pacha, Pacha à Trois Queues*, Paris 1953, 143-145.

9. Letter dated 15 June 1729, Archives du Ministère des Affaires Etrangères, Correspondance Politique, Turquie (CP Turquie) vol. 81, 40-44.

10. M. Arif, " Humbaraci Ahmed Pasha [Bonneval] ", *Tarih-i Osmanî Encümeni Mecmuasi (TOEM)*, vol. 3/18, 1328, Istanbul, 1143-1144.

11. S. Gorseix, *Bonneval Pacha*, 160.

above-mentioned Bombardier Corps[12].

Two years later, three experts were summoned from France to serve under Bonneval's command. They arrived on January 17, 1734 in Istanbul and were introduced to the Grand Vizier. They greeted the Grand Vizier in Muslim attire and were granted 200 *kurush* each. Later, all three — the Marquis de Mornai (of French origin), Comte Ramsay (of Scottish origin) and L'Abbé Macarthy (of Irish origin) — embraced Islam and found positions in the Ottoman state[13].

Bonneval Ahmed, together with the three new European officers, initiated a training program for 300 young men brought from Bosnia to be trained as bombardiers in the newly built barracks at *Ayazma* Palace in Dogancilar, Üsküdar, Istanbul. These young men were selected to be prepared as soldiers with expertise in mathematics. This new Corps of Bombardiers would constitute a model for the reorganization of the entire Ottoman Army. It was the first time that an Ottoman corps was trained according to European methods under the command of a European officer.

These new bombardiers were different from the classical bombardiers in two important ways : first, while the classical bombardiers were employed in different parts of the country, the members of this new corps were gathered and trained in the capital of the Empire. Second, bombardiers in the new corps were paid regular salaries instead of obtaining fiefs, and underwent practical and theoretical training by instructors who were specialists in mathematics and military techniques.

By an imperial decree dated January 25, 1735 confirming the establishment of the Bombardier Corps, a regulation book was issued setting the terms of the Corps. By this regulation the Bombardier Corps was officially reorganized and put under the command of Bonneval Ahmed. First, officers were chosen, their salaries determined, and a hierarchy among them was established. The Bombardier Corps was directly attached to the office of the Grand Vizier ; its financial administration would be carried out by a specially appointed treasurer. The Corps would be headed by the *Humbaracibashi* (Chief Bombardier), carrying the title of *Alaybashi* (Commander-in chief), who had the right to appoint and depose the officers[14].

The Corps was divided into three *odas* (divisions), each consisting of 25 *zabits* (officers) and 75 *nefers* (soldiers).

12. CP Turquie, Supplément vol. 13, p. 150v-151, A former Turkish ambassador to France Salih Münir Pasha, mentions this report ; Salih Münir, " Bonneval Pacha, Son Influence sur les Relations Extérieures de la Turquie ", *Revue d'Histoire Diplomatique*, v, xxi, 378-393.

13. Journal de l'Ambassade de M. Le de Marquis de Villeneuve, Archives de Nantes, CP Turquie supp.vol. 17, 203.

14. *Basbakanlik Osmanli Arsivi* (BOA) *Maliyeden Müdevver Defterler* (M.MD), n° 5941, p. 4-7 ; M. Sami, *Tarih-i Sami ve Sakir ve Suphi,* Istanbul, Rasid and Vasif Efendi Press, 1198H, p. 58a-59a; I. Hakki Uzunçarsili, *Kapikulu Ocaklari*..., II, 122-125.

N°	First Division	Second Division	Third Division	Daily salary (each person)
1	Chief Sergeant			240
3	Leader of the division	Leader of the division	Leader of the division	240
6	2 Leaders of Fifty	2 Leaders of Fifty	2 Leaders of Fifty	90
9	3 Leaders of Thirty	3 Leaders of Thirty	3 Leaders of Thirty	50
3	Secretary	Secretary	Secretary	90
3	Teacher	Teacher	Teacher	40
3	Sergeant	Sergeant	Sergeant	40
6	2 Drummers	2 Drummers	2 Drummers	24
30	10 Corporals	10 Corporals	10 Corporals	30
1	Second Sergeant			60
1	Imam of the Corps			40
1	Doctor of the division			60
1		Engineering teacher		60
1		Drawing teacher		60
1		Chief Surgeon		60
2		Second Surgeon	Third Surgeon	36
1			Surgeon	36
1			Religious leader	40
1			Steward	50
75	**25 officers**	**25 officers**	**25 officers**	**Total salary**
	1498 Akçes	1255 Akçes	1209 Akçes	3962 Akçes

Looking at the above table, we can see that officers with administrative duties were gathered in the first division, those in charge of education were in the second division, and officers taking care of fiscal matters were found in the third division. In the light of the fact that in each division there were seventy-five soldiers and one officer for every three soldiers, we can conclude that there was a well-defined system of organisation and training within the Bombardier Corps.

TEACHING AND TRAINING IN THE BOMBARDIER CORPS

Bonneval Ahmed Pasha took a great personal interest in the training of the bombardiers from the first day of the Corps' establishment. He crossed the Bosphorus to Üsküdar every day in a rowboat, inspected the training given in the barracks, and taught the soldiers new bombarding techniques. He designed a new type of mortar which was much more economical in gunpowder consumption (see figure 1).

The Chief Sergeant of the corps was Engineer Selim, who was also in charge of practical training. Engineer Selim was appointed to his post in 1735 with a daily salary of 240 *akçes*. Known as *Cenk Mimarbashi* (Chief Architect of War) he had studied war techniques in Europe. He taught his students European military techniques (such as the construction of fortresses and trenches using geometrical calculations, technical drawing of military architecture, and design) in the Corps of Bombardiers between 1735-1741[15].

As for theoretical training, there were several instructors. Haci Mahmud Efendizade Mehmed Said Efendi, the son of *Mufti* of Yenishehir, taught geometry in the corps. Said Efendi was *hodja-i mühendis* (engineering teacher) in the second division and received a daily salary of 60 *akçes*[16]. While he taught geometry *(hendese)* in the Bombardier Corps, he wrote monographs about the usage and the properties of tools used to measure artillery ranges. Among them a treatise (1737) where he described the *Dürbünlü rub'u müceyyebi zü'l-kavseyn,* a special tool he invented to determine the ranges of mortars[17]. Based on the Thales Theorem, it was used for calculating the 3rd angle and two sides of a triangle when the two angles and one side were known (see picture 2). His book on mathematics and geometry was called *Risale fi'l-Hendese ve'l-Hesab* including geometrical figures at the end[18]. On the other hand, Said Efendi had four books on astronomy[19].

The other instructor of the Bombardier Corps were Ali Ahmed Hodja of Kasimpasha[20], Süleyman b. Hasan of Istanbul[21], and *hodja-i muallim* Osman b. Abdullah[22]. Technical drawing was taught by Ibrahim of Istanbul until 1739[23].

After the death of Bonneval Ahmed Pasha in 1747, the Bombardier Corps was administered by his adopted child, Süleyman Aga. However, the corps gradually lost its efficiency and close down after a while[24].

Thus, we have seen that the reform movement in the Ottoman army was initiated with the establishment of the Corps of Salaried Bombardiers. As a result

15. BOA, M.MD, n° 5941, p. 48. It is highly possible that Engineer Selim is the same person as the Marquis de Mornai, who was summoned from France to help Bonneval.

16. BOA, M.MD, n° 5941, p. 48.

17. Library of Topkapi Palace Museum (TSMK), Hazine, n° 1753/3, ff. 18b-35b.

18. Cairo Dar al- Kutub (Egypt) n° 4773, 15 ff. with 20 figures.

19. *Cadvalu Sa'ati Matali' al Baladiyya li 'Arzin Mâ,* Library of Süleymaniye (Lib.S), Esad Efendi, n° 2055, 75b, copy in 1770. *Risala fi'l-Fark Bayân al-Sa'at al Zavali va'l- Sa'at al Gurûbî,* (Lib.S), Esad Efendi, n° 3074/11, ff. 58b-59a ; *Risala fi Rasm davair al Falakiyya fi Rub'u al Mukantara,* Lib. TMSK, Hazine 1753/11, ff 72b-73b, copied by the author. In Arabic *Risala fi'l-Rub'u al Shikazî* Said Efendi tells about the features and usage of a tool called " rub'u Shikaze ", Library of Nurosmanniye, n° 2918/4, ff. 46b-59b.

20. His name is mentioned in the second order of the list of the Bombardiers Corps in 1739, BOA, M.MD, n° 5941, p. 48.

21. His name is mentioned among the officers in the inspection book, 1739, BOA, M.MD, n° 5941, p. 42.

22. *Hodja* in the first division of the Bombardier Corps in 1735. BOA, M.MD, n° 5941, p. 12.

23. His name is mentioned among the officers in the inspection book, 1739, BOA, M.MD, n° 5941, p. 42, and see marginal note BOA, M.MD, n° 5941, p. 110.

24. M. Kaçar, " Osmanli Imparatorlugunda Askeri Sahada Yenilesme Döneminin Baslangici ", F. Günergun (ed.), *Osmanli Bilimi Arastirmalari (Studies in Ottoman Science)*, Istanbul, Istanbul University, 1995, 209-225.

of this change, candidate bombardiers were recruited and gathered in a central barracks in Istanbul, where they were trained in mathematics and the art of using mortars. (The fact that they were paid regular salaries was an important change in itself). Thus, Ottoman military reforms were aimed at improving the technical training of soldiers rather than altering the organisation of the army itself. The training carried out in this corps was viewed primarily as training provided in the context of military service. This kind of training should be distinguished from the mathematical training given in independent schools that will be discussed in the coming pages.

THE FOUNDATION OF THE ENGINEERING SCHOOL AND THE BEGINNING OF AN INDEPENDENT MILITARY ENGINEERING EDUCATION

The first attempt to establish an independent institution providing military technical education was undertaken while Gazi Hasan Pasha was the head of the Ottoman Navy. Gazi Hasan Pasha had visited European ports and dockyards and was aware of European technological developments.

Baron François de Tott, a French nobleman, entered the Ottoman service and served for six years in high government offices (without becoming a Muslim). He was charged with several military and technical projects. He first served at the fortifications near the Dardanelles Straits where he tried to introduce a European style canon foundry and European artillery techniques to the Ottoman army. Within this framework, he was influential in the foundation of *Sürat Topçulari Ocagi* (Corps of Diligence) upon the order of Sultan Mustafa III in 1774[25].

Baron de Tott and Sr. Kermovan, another French expert, were charged by Gazi Hasan Pasha with establishing a class on geometry in the Imperial Arsenal. This class called " Hendesehane " or " Ecole de Théorie et de Mathématiques " opened with ten students on April 29, 1775[26].

The *Hendesehane* was originally a school where different branches of mathematics were taught to students coming from various military corps — including artillery, fortification, and the navy. It was different from the above-mentioned Corps of Bombardiers, where mathematics was taught as an integral part of the bombardier training[27]. Baron de Tott, Kermovan, and Campbell Mustafa (of Scottish origin) lectured mathematics in the *Hendesehane* for

25. M. Kaçar, " Osmanli Imparatorlugu'nda Askeri Teknik Egitimde Modernlesme Çalismalari ve Mühendishanelerin Kurulusu ", F. Günergun (ed.), *Osmanli Bilimi Arastirmalari II* (*Studies in Ottoman Science II*), Istanbul, Istanbul University, 1998, 79-81.

26. The report of the French ambassador dated 3 May 1775 : " L'école de Théorie a été ouverte le 29 Avril (1775) dernier à l'Arsenal sous la direction du Sr. Kermovan et d'un renégat anglais nommé Mustafa Aga avec la surveillance de M. de Tott ", CP Turquie vol. 161, p. 171r.

27. M. Kaçar, *Osmanli Bilimi Arastirmalari II*, 82-83.

about four months until September, 1775[28].

In 1776, a new regulation was issued by Grand Admiral Gazi Hasan Pasha, and the *Hendesehane* was reorganised in line with the classical Ottoman bureaucratic and financial structure. The teaching staff included a professor (*hodja*), an assistant professor (*halife*) and a keeper of instruments. There were about ten students (*shakird*) in the *Hendesehane*. Foreign specialists could also teach in this institution. After Baron de Tott left Istanbul, Seyyid Hasan Hodja, the Second Captain of the imperial fleet and a member of the *ulema*, was appointed director and professor to the school. Thus, the *Hendesehane* established under the administration of European specialists, continued to give instruction through the contributions of native teachers[29].

From 1781 on, the *Hendesehane* was also called *Mühendishane* (School of Engineering). During Halil Hamid Pasha's term as Grand Vizier (1782-1785), the Ottomans were on good terms with France[30]. At the request of the Ottoman state, Lafitte-Clavé and Monnier, two French military experts, were sent to Istanbul and put in charge of the reform projects in the Ottoman army. They were also asked to strengthen the fortifications at the Empire s borders, and to reorganise the training in the Maritime Arsenal and the School of Engineering as well. Thus these French experts started to train officers in modern military arts and sciences (artillery, navigation, and fortification) together with teachers from the Ottoman *ulema*. After 1788, when the Frenchmen left the Empire[31], the native *medrese* teachers took over the instruction in the imperial schools of engineering, where the classical Ottoman science books were used alongside European books until the end of the eighteenth century.

With the reign of Sultan Selim III (1789-1808) began a military reform movement called the *Nizam-i Cedid* (New Order). Within the framework of this reform, new regulations were set for the imperial schools of engineering. Accordingly, the *Mühendishane-i Cedid* (New *Mühendishane*) was established in 1793 to train those military corps that required mathematical and technical knowledge (i.e. canoniers, bombardiers, and miners). The Ottoman engineer-teachers, who had been the students of French engineers such as Lafitte-Clavé and Monnier and later taught fortification in the Arsenal in 1784, now lectured in this new *Mühendishane*. The teaching staff included a professor (*hodja*) and four assistant professors (*halife*), who instructed the bombardiers and miners in geometry, trigonometry, altimetry, and land surveying.

Between 1801-1802, about one hundred students selected from the corps of bombardiers, sappers, miners, and architects were recruited into the

28. W. Eton, *A. Survey of the Turkish Empire*, London 1789, 73-74n.

29. BOA, M.MD, n° 8882, p. 82.

30. A. Toderini, *De la Littérature des Turcs*, vol. 1, Paris, 1789, 161-162.

31. F. Hitzél, Défense de la Place Turque d'Oczakow par un Officier du Génie Français (1787), *II. Tarih Boyunca Karadeniz Kongresi*, Samsun, 1990, 648.

Mühendishane to be trained as engineers. The teaching staff included a professor and five assistant professors. A regulation issued in 1806 betrayed the influence of both the *medrese* and European educational systems. The school was organized into four classes with four teachers. This regulation also introduced the practice of graduating from one class to the other, enabling the students to pass to a higher rank. This was a novelty in Ottoman educational life[32].

Thus, with the establishment of the *Mühendishanes*, an Ottoman-European synthesis started to gain ground in Ottoman educational life. As a result of the research conducted so far, it is clear that modern science and technology were introduced to the Ottoman state at the demand and through the efforts of Ottoman scholars and administrators. Experts recruited from Europe contributed mainly to the teaching of new techniques.

FIGURES

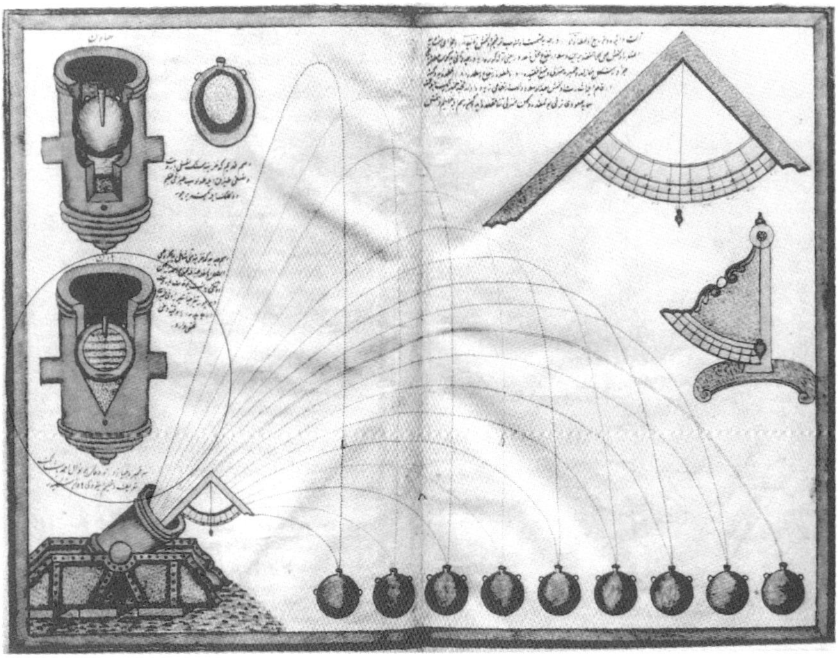

1. The new type of mortar as designed and used in the drills of the Corps by Comte de Bonneval (Ahmed Pasha), Chief of the Bombardier Corps (see the figure within the circle).

32. M. Kaçar, *Osmanli Bilimi Arastirmalari II*, 87-114.

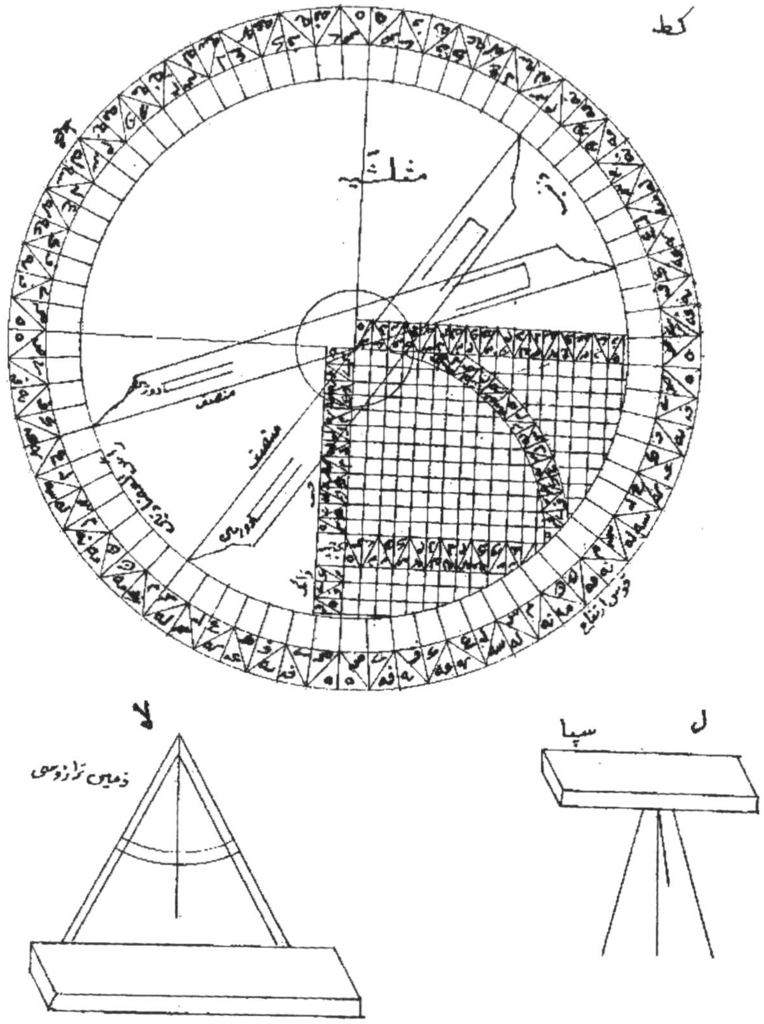

2. The drawing of instrument called " Müsellesiye " invented by Said Efendi, Ottoman Engineer in 1737.

SCIENTIFIC MOTIVATION FOR AND MOOD FROM THE EXPERIENCE OF THE EGYPTIAN EXPEDITION

Jean DHOMBRES

The historiography of the Egyptian expedition is as monumental as the wealth of the Egyptian civilisation it revealed to an astonished European world. The expedition is one of the most studied episodes in French history. It is even difficult to find the name of an intellectual and member of the expedition who did not leave letters, memoirs, or thoughts. Precisely from an intellectual point of view, this expedition raised fascinating historical questions[1]. Such questions concern, for example, the conditions for the birth of Egyptology, and how the joint work was organised which gave rise to the volumes of the *Description de l'Egypte*, or *Recueil des observations et des recherches qui ont été faites en Egypte pendant l'expédition française*. Such a collection of so many pages played for the new century the part played earlier by the *Encyclopédie*[2]. Another question is the way ideas of colonisation were coined, based on scientific achievements of the Enlightenment and what brought about not only a sense of a supremacy of European thought, but also its demonstration. Not to mention the strange association between a campaign issuing from the Revolution and the founders and heirs of the new scientific movement, characterised by the *Ecole polytechnique* (as the German mathematician Felix Klein strongly emphasised at the beginning of the present century). While studying how the French scientific community behaved during the early nineteenth century, I

1. A large bibliography about the expedition was already provided by H. Munier in 1943. The late scholar J.E. Goby provided valuable bibliographies on the expedition as well, and particularly on its scientific aspects, Premier Institut d'Egypte. Restitution des comptes rendus des séances, *Mém. Acad. Insc. Belles-Lettres*, nouvelle série, t. VII, 1987. More books were devoted to bibliographical references and testimonies, like P. De Meulenaere, *Bibliographie raisonnée des témoignages oculaires imprimés de l'expédition d'Egypte*, Paris, F. et R. Chamonal, 1993. Recent bibliographies can be found in J.N. Brégeon, *L'Egypte française au jour le jour, 1798-1801*, Perrin, 1991 and P. Bret, *L'Egypte au temps de l'expédition de Bonaparte, 1789-1801*, Paris, Hachette, 1998 (La vie quotidienne).

2. All references to the *Description de l'Egypte* will be to the second edition, Paris, Panckoucke, 1821-1829, 26 vol. in-8°, 11 vol. in-plano. Referred here in an abridged form : D.E.

wished to encompass all such questions under the generic name of the " Egyptian laboratory "[3]. I certainly do not aim here at describing the various activities of this " laboratory " once more, even in this interesting symposium devoted to the Ottoman Empire. We need only remember that this Empire was violated by the French army, but eventually and rather quickly succeeded in expelling the invaders. Did it succeed well in throwing away all " modern " ideas brought by French intellectuals ? Was it to forget the project of an independent Egypt recovering its bright scientific past ?

My aim, however, is not to look for Egyptian and Ottoman reactions to the invasion : recent studies fortunately brought to light very interesting descriptions of such reactions such the present symposium explores[4]. But another kind of reaction of scientists and/or engineers to Egypt may be of some interest. Without playing with words, I should add that there was a reaction to two countries : Ancient Egypt and Ottoman Egypt. Both countries are certainly linked by the passing of time in the *Description de l'Egypte* ; which devotes as many volumes to the first as to the second. Modern Egypt should help one to understand the still indecipherable Antique Egypt, and ancient Egypt may provide the conditions for a regenerated Egypt, i.e. a colonised one. Officially, such were the two main reasons given for an astonishing enterprise, ruled by engineers ; it finally became not really Archaeology, but something else. It was to be called Egyptology ; the word encompasses not only one science, but is the focus of different sciences, in the same way as the word engineer was soon to represent a focus for many kinds of scientific specialisation's[5]. To keep an alert mind we must not forget that during the 18th century British armies never tried to undertake a History of India, nor did the Dutch colonisers of Java give extensive descriptions of Borobudur.

What made scientists and engineers react so strongly to Egypt ? What gave them the desire to undertake both a *Description* and an exploration of Egypt ? These are questions I wish to investigate by providing a sort of reconstruction : I will try to recover the intellectual motivations of a small and often disoriented community[6]. I will also try to show that something purely scientific was at stake. I do not forget that the present symposium takes place during the International Congress for History of Science. In fact, through the experience of

3. See J. et N. Dhombres, *Naissance d'un pouvoir. Sciences et savants en France (1793-1824)*, Paris, Payot, 1989, chap. 2 : *Le laboratoire d'Egypte.*

4. I particularly refer to H. Laurens, *Les origines intellectuelles de l'expédition d'Egypte. L'orientalisme islamisant en France (1698-1798)*, Istambul, Paris, Isis, 1987, and H. Laurens, *L'Expédition d'Egypte*, 1798-1801, Paris, réédition Le Seuil, 1997.

5. J. Leclant, " La modification d'un regard (1787-1826) ", *Comptes rendus Acad. Insc. Belles-Lettres*, Paris, 27 novembre 1987.

6. Most historians, once they were able to get rid of the imperial and epic tone, noticed the alternance of periods of depression and periods of enthusiasm among scientists in Egypt. A thorough study of letters could lead to an interpretation about the sense of supremacy so often shown by the colonisers in Egypt.

Egypt, I wish to understand how " positive " science also emerged from the transformation of " conquering " and the thoroughly explicative science of the Enlightenment[7].

In brief, Egypt provided the ground for discussions, debates, and even conflicts about what science achieves. This debate is what constitutes the historical object of this short paper.

Even if I can only be but sketchy, in order to recover a mentality spread among a small group, less than two hundred persons[8], I first have to mention two related processes, which were caused by the conquest and will mould its posterity.

PROGRESS IS NOT IRRESISTIBLE : *LES LUMIÈRES* ARE NOT SUFFICIENT FOR A SUCCESSFUL COLONISATION

The question of the transfer of the Enlightenment haunted many minds in Egypt. An anecdote is revealing. In the belief they were delivering an efficient propaganda, French engineers wished to launch a balloon over Cairo, precisely on the anniversary date of the foundation of the Republic, and only some weeks after their arrival. From June 1783 onwards, in so many cities in Europe, such balloon launchings always attracted crowds, admiration and enthusiasm towards a new era of mankind. In Cairo, so wrote so many observers, inhabitants went to their own business without caring to look over their heads[9]. Were Egyptians not ready for Enlightenment ? This question will organise the main part of Fourier's analysis of the experience issued form the expedition. His text was published in 1810 as an introduction to the *Description de l'Egypte*, under the characteristic name of *Préface historique*. The purpose was to explain why an army, and more surprisingly a defeated one, so intensively investigated a country and its history. A glorious but " singular " aspect, so claimed Fourier. Even under the yoke of an Emperor, French intellectuals required rational reasons for political actions, and Fourier never changed his opinion about Egypt, " After a long exile, sciences again see their homeland and are ready to shape it far better "[10].

7. J. Dhombres, " La querelle des zodiaques ", in *Douze leçons d'histoire des sciences* (to appear).

8. This number adds the members of the Commission des sciences et des arts to those belonging to the Cairo Institute, and some officers (Y. Laissus, *L'Egypte, une aventure savante*, Paris, Fayard, 1998). I appreciate the scientist mood among historians who wrote there were approximately 187 members (plus or minus 0.5 perhaps !).

9. See for example the testimony of E. Jomard, *Souvenirs sur Gaspard Monge et ses rapports avec Napoléon*, Paris, E. Thurot, 1853, 36 and P. Bret, *op. cit.*, 170-172. For balloons in general around this time, see C.C. Gillispie, *The Montgolfier brothers and the invention of aviation, with a word on the importance of ballooning for the science of heat and the art of building railroad*, Princeton Univ. Press, 1983.

10. J. Fourier, " Préface historique ", D.E., t. 1.

Thus the observed indifference of Egyptians to " modernity ", (which is not the same as resistance or opposition), provided an acceptable answer for the failure of French colonisation. We certainly cannot forget that such an answer was pleasant to Napoléon, who was so eager not to appear as a loser, or even worse, as a runaway. But it is far too simple to assume that Fourier just acted as a courtier : the whole life of this man, in the sciences as well as in politics, shows he was able to neatly state his thoughts, using the convenient amount of diplomatic rhetoric[11]. For Fourier, and he was not the only one to think so, a serious and deep consequence of this expedition was to show that science, progress, and reason, were certainly not enough to manage a country. As many others, he took for granted the minority level of Egyptians, compared to Westerners. But it was not enough to claim, as Kant did, *Sapere aude*, in order to leave the minority level. Fourier tells in a very explicit way in the *Préface historique* that there also must be the will from the majority of human beings. " Egyptians are unable to judge the necessary relations, which unite speculative science, technical arts, industrial progress, treasury management, and justice management, that are all the elements of a regular government, able to warrant happiness and prosperity to the people as a whole "[12].

Contrary to what Condorcet thought, and to what had been an implicit ideology in order to end the Revolution during the Directoire period, progress was not irresistible because it was not a universal value.

THE " COLD " LOOK OF OBSERVERS

In Egypt, even if it is in an ironical way, what the mathematician Gaspard Monge straightforwardly wrote in a letter to his wife in order to transform Egypt, duly expressed his idea on the western supremacy and the role of progress. " When this country would have been inhabited, planted and pierced for fifty years, then it would become a terrestrial paradise "[13]. Egyptians do not exist !

They do not even exist in the *Description*, in the sense that their opinion is not required. This attitude adopted a particular twist. The purpose of a systematic and scientific survey of the land fertilised by the Nile was to transform inhabitants into mere objects, countable as well as numerous hieroglyphs engraved on columns and friezes. The analytical spirit is quite clear under the

11. See J. Herivel, *Fourier, the Man and the Physicist*, Oxford, Clarendon Press, 1975, or J. Dhombres, J.B. Robert, *Fourier, créateur de la physique mathématique*, Paris, Belin, 1998. Among historians of science and due to some kind of prejudice, there is a lack of investigation about political behaviours of scientists.

12. J. Fourier, *Discours de réception à l'Académie française* (17 avril 1827), Paris, Didot, 1827.

13. " Lorsque ce pays-ci aura été habité, bâti, planté, percé pendant cinquante ans ; ce sera un paradis terrestre ", a letter dated 14 November 1798. See R. Taton and *alii, Correspondance d'Italie et d'Egypte de Gaspard Monge* (to appear).

pen of F. de Rozière, an engineer for mining : " No other country provides to the eye such a dependence form a highly developed society, under a unique bunch of physical factors which could alone be studied "[14]. Egyptians only existed to ascertain an older civilisation. As so little was known about it, Egypt thus took the role of a gigantic laboratory, where cartographers, astronomers, and engineers might measure in complete freedom. Through topographical statistics, as began the fashion in France as well, doctors were measuring their toll of deaths in hospitals. We may say that in the opinion of many scientists and engineers, Egypt belonged to science because the enquiry was to be made public, i.e. universal. In other words, but our modern words, Egypt thoroughly studied might help to create the universal model for westernisation. Therefore, as a consequence, scientists in Egypt thought they had to adopt a " cold " look, what they thought was the scientific look[15]. Many hoped that in the long range, it was for the benefit of Egyptians themselves. But the political, economical and social consequences issued from the scientific survey could only lead to an imposition from a non-Egyptian world.

An Egyptian " theorem " and reaction towards a positive history

An exhibition of conquering science was also taking place in Paris just some weeks after Aboukir disaster. It took the form of a meeting of an international congress ; it is the first known one. It was organised to verify all measurements and computations, which were to lead to a new metrology — the so-called decimal metric system[16]. All such academicians, Lalande, Delambre, Laplace, were celebrating the meeting of so many scientists joining their efforts in order to define the metre. As its length was explicitly linked to the size of the Earth, the metre could be viewed as a universal measure, and therefore compulsory for all. The metre became a monument, symbolising and uniting both the Enlightenment and the Revolution through the active role of scientists.

An analogy thus spread among officers and engineers in Egypt, scientists as well, who were following with due attention everything which took place in France. This analogy almost took the form of a theorem, the " proof " of which we may summarise in the following way. There exists a language for Nature, the name of which is science ; men are able to read it even if the whole geographic world has not been explored ; the metre is doubly a monument. It tells

14. F.M. de Rozière, " De la constitution de l'Egypte et de ses rapports avec les anciennes constitutions de cette contrée ", D.E., t. 20, 211.

15. Officers did have the feeling to deal with partners in arms, and they were right as finally they were to lose. But scientists almost never mentioned Egyptian intellectuals with whom they could have had any discussion on some basis of equality.

16. J. Dhombres, " Mesure pour mesure, universel contre régional : le système métrique comme action révolutionnaire ", in A. Jourdan, J. Leerssen (éds), *Remous révolutionnaires : République batave, armées françaises*, Amsterdam, 1996, 159-199.

the measure for the ten-millionth part of the quarter of a meridian of the ellipsoid which is flattened around the two poles — the Earth. But is also tells the wealth of the analytic science which presided to this measurement ; the metre is a warrant to the principles of Mechanics and therefore to all its consequences. Therefore, in Egypt, where hieroglyphs cannot be read, the bright past civilisation has to be read through its grandiose monuments. A careful reading is necessary. In the same way as the metre, pyramids are double monuments. In one sense as it defies Time, but also a monument as a geometry textbook. Edme Jomard, an engineer who will become director for the *Description de l'Egypte*, explained how science could be read in the monument itself. " In a pyramid, the comparison of the basis to the apothem could be given in the excess of one to the other, the side of the *aroure*, a measure of one hundred cubits, a quarter of the Egyptian *stadia*, the basic element of every day life for all agrarian measures… It showed then a way to find again, at any time, the side of the *aroure*, and consequently of all measures ".

Egyptian science, as any science, overcame time through the universal language of measurements. Jomard thus was able to formulate the Egyptian " theorem ". " From proportions, which are shining from these monuments, one may deduce rules along which they were built ; thus, as they are the fruits of Egyptian science, they must keep its basic elements, and it cannot be impossible to discover such elements through monuments ".

" O my friends, you are so lucky to go and see Thebes ", so claimed Gaspard Monge, sincerely moved at the very moment he was quickly leaving Cairo, and the Institute, in order to fly towards Paris with the general in chief. We have to keep in mind an enthusiasm for the discovery of a grand civilisation, which had been allowed by a grandiose science. Its force may be measured by the heights of Egyptian monuments, and due to their very precision, geometric drawings as well as perspective drawings might show how strictly right are monuments. In spite of all hidden corridors, rectilinearity could not but have an astronomical meaning. It was not necessary to think any more ; it was no longer necessary to use the traditional language of styles which architect could provide, or the also traditional language of historians. Alone, science could make History. This was paving the way to scientism.

In fact, Joseph Fourier, as secretary of the Egyptian Institute in Cairo, reached Thebes two months later, chief of one of the " literary " expedition sent to Higher Egypt. As a specialist of mechanics — this was the only subject upon which the young professor of analysis at the young *Ecole polytechnique* had already published a year before — Fourier should have been the right man able to " see " in the same way what Egyptian science could tell. On the contrary, he exhibited a real sense of suspicion. Which underlines a change in the mood of conquering science ; precaution was thought necessary, and older kind of knowledge should not as easily be replaced. Rigour, intellectual rigour, was now the main effort of knowledge and this very rigour was to avoid scientific reasoning alone when it could not ascertain truth. Fourier will later summarise by using the word " exact " : " the objective of my researches is to discover the exact consequences, which can be deduced from these precious remains of an old astronomy ". Exactness here has to be understood in a somewhat restrictive way : science achieves something, but not everything.

Sculptures had indeed been found in Higher Egypt, which were viewed as zodiacs, i.e. representations of all different positions of the Sun along the year, divided into twelve periods ; such regions had been located and named according to the diverse animal forms taken by constellations. Could such calendars yield " positive " or scientific information concerning the dates when Egyptian monuments were built ? Such was the question[17].

An affirmative answer was a direct consequence of the Egyptian " theorem ". Which Fourier first accepted in this occurrence. He did so because historians were discordant on the subject of chronology. Had not Larcher, the most recent learned translator of Herodotus, claimed in 1786 that " Egyptian chronology was the thorniest problem of antique history "[18] ? Boldly, on the site of Esne, using as in an academic paper the pronoun " we ", Fourier wrote in his personal diary : " We arrived at Esné on the 22nd in the morning, and visited the temple which is in the city... We have seen on the ceiling important and well preserved sculptures showing zodiacal signs. We will publish a specific paper about interpretation of these sculptures ; it will help us to assign the period of time when the sphere, to which the zodiac is but a part, was established. It goes back to four thousand and six hundred years before Christ ; and the date for the building of the temple cannot differ from more than... centuries[19]. By the way, this second hypothesis is rather uncertain, but the first one cannot be contested "[20]. Clearly, Fourier thought that pure astronomy, using precessions of the equinox, might provide a precise date for the temple. Obvi-

17. There were in particular two zodiacs found in the temple of Denderah, one rectangular, and the other circular. In spite of the fact we only know now, that its superstructures are recent (first century AD.), or perhaps because copies are more impressive at first glance, this temple has inspired most engineers in Egypt, and painters as well like Denon. It is there that they gave up their Greek look, in order to see Egypt as such. The zodiacs also played an important part, a stamp of scientificity. There has been a huge literature devoted to such astronomical sculptures, almost to the end of the nineteenth century. The circular zodiac was even stolen and brought to Paris in 1822, where it can still be seen in the Louvre. It is interesting to notice that no mention is made of the quarrel it created, in particular from science against religion.

18. P.H. Larcher, *Histoire d'Hérodote, traduite du grec, avec des remarques historiques et critiques, un essai sur la chronologie d'Hérodote, et une table géographique*, Paris, Musier/Nyon, 1786/1787.

19. Fourier left this passage without numbers. He certainly wished to later compute some approximation on the possible number of centuries elapsed since the building of the zodiac, depending on the precise reading of the zodiac (he did not have the drawing at the time he wrote his diary), or on some further observations. The fact that he did not complete his task symbolically reveals his change of mind, and caracteristically expresses the kind of censure he exerted in Egypt : silence was considered as better than unproven interpretation.

20. Extrait des notes de voyage de J. Fourier, *Bibliothèque égyptienne*, G. Maspéro (éd.), vol. 6, Paris, 1904, 180.

ously, he was aiming at showing that biblical dates were irrelevant, and that the history of the Earth was far greater than what biblical sources tended to say. Indeed, all his efforts will later be to correct such an assertive sentence, he will have thus to correct the " we " he used ; he will have to teach a lesson to all his colleagues in Egypt[21]. He later will write : " The way signs of the zodiac were designed rather belong to religion than to science : it even may have its origin in an older state of astronomical knowledge when twelve signs of thirty degrees were not considered, but twelve unequally located constellations "[22]. Further, Fourier explains that astronomy may help History, but not replace it.

E. Jomard, an engineer specialised in geography, later to become a member of the *Académie des Inscriptions et Belles-Lettres* maintained that he kept a serious look upon astronomical fragments " because they offered a basis for certainty, and as having common points with ideas and notions of modern times ". And it is against Jomard that J.F. Champollion will show, by reading first names, that indeed zodiacs were built around the beginning of the Christian era. But he was not the only author to provide a convincing proof. A scholar and not a scientist, Letronne, just earlier had shown the same through a careful reading of Greek inscriptions on walls of the temple. I am wrong not to call Letronne a scientist ; in fact by giving foundations to epigraphy, he had adopted in history a scientific way.

This transformation of Fourier's mind specifically represents the passage to positivism, and it is noticeable that it had to travel from scientism. What we call scientism, which remained in minds like that of Jomard, really had to be fought : positivism did not arrive quietly. Two arguments were provided. The first one, which was to take time to reach historians, was to pretend that we should not scientifically deduce facts form antique texts, even from zodiacs, by presupposing a kind of scientific knowledge not transparent to the authors of the text. In other words, there is not in any scientific text something useful for us, which was not already clear to the ancient authors of the text[23]. The second one is clearly linked to the first, and is related to a universal definition of science : science cannot be but public ; it cannot be hidden, or resulting in a

21. J. Dhombres, " Joseph Fourier égyptologue ? Une pédagogie sévère dans l'exercice de l'autorité intellectuelle au début du XIXe siècle ", in *Bicentenaire de l'expédition d'Egypte. Les Bourguignons et le Levant*, Auxerre (to appear).

22. J. Fourier, " Recherches sur les sciences et le gouvernement de l'Egypte ", D.E., t. 9, 24. Later, J.B. Biot modified the " scientific " interpretation of zodiacs to better adapt to this rule of adequacy between a scientific knowledge and its use for history. He will deduce a wrong date for the zodiac, around 700 BC. A numerically large literature has been devoted during the first half of the 19th century to " astronomical " datations by zodiacs ; it offers a good place to see the building of the mentality of professionnal science.

23. Jomard, by reading some geometrical knowledge directly in the Egyptian pyramids, presupposed that Egyptian mastered all geometrical proprieties, and wanted to exhibit such properties, not by a text but by a monument. He was reading in a modern way a pyramid, and presupposed there was an Egyptian message. Jomard's kind of reasoning is maintained today in so many books on Egypt, with the so-called mysteries of the pyramids. The word mystery is not refering to our lack of knowledge, but to a surplus of knowledge attributed to Egyptians of the earliest dynasty.

sort of an initiation. If there was some scientific knowledge in Thebes, we cannot recover it from the myth of some mysteries of Isis. Fourier went against a tradition of the thought even in the eighteenth century, which was so beautifully exhibited by Baltrusaitis[24].

That Fourier helped the then young Champollion in Grenoble, and later in 1822 admired his work, should not astonish us. He did it, in spite of the fact that Champollion's result totally abolish Fourier's thesis on the zodiacs. But in fact, by better means, it confirmed the dates which Fourier assigned for the time of the " splendour of Thebes ". He then could claim that Champollion was doing " geometry ". Including all limitations Pascal gave to the expression, I have tried to show the conditions for the birth of such an *esprit de géométrie* during the Egyptian expedition.

24. J. Baltrusaitis, *La quête d'Isis. Introduction à l'égyptomanie*, Paris, O. Perrin, 1967 (nouvelle édition, Paris, Flammarion, 1997).

L'EXPÉRIENCE PRÉALABLE DE L'EMPIRE OTTOMAN DANS LA COMMISSION DES SCIENCES ET ARTS DE L'EXPÉDITION DE BONAPARTE EN ÉGYPTE (1798-1801)

Patrice BRET[1]

L'expédition de Bonaparte en Egypte constitue une parenthèse — mais une parenthèse importante — dans l'histoire de l'Egypte et de son appartenance à l'Empire ottoman. Lorsque les troupes de Bonaparte débarquent à Alexandrie, le 1er juillet 1798, elles arrivent dans une province partiellement dissidente de cet empire, et sous le contrôle des Mamelouks Murâd bey et Ibrâhîm bey. Le but déclaré de l'expédition française est justement de libérer l'Egypte du joug mamelouk, au nom du Sultan, allié traditionnel de la France depuis deux siècles et demi : Bonaparte se place ainsi implicitement comme un successeur du Capitan pacha, Hasan pacha, qui avait rétabli une première fois le pouvoir ottoman douze ans plus tôt[2].

Les enjeux réels de l'expédition, géopolitiques, économiques et scientifiques étaient sans doute tout autres. Il convient de les replacer aussi dans le contexte intellectuel français du temps, qui proclame l'unité du savoir humain avec le mouvement encyclopédiste et qui, dans le droit fil de la Révolution française, envisage la régénération de l'homme — et en l'occurrence de l'Egypte — par la science et l'éducation. Retrouver les racines de la civilisation, régénérer l'Egypte, patrie d'origine des sciences et des arts, et en faire l'inventaire systématique : voici quelques thèmes forts de l'expédition qu'il ne faut pas considérer comme simple rhétorique, et qui ne sont pas alors incompatibles avec un projet colonial parallèle — tirer de l'Egypte des productions coloniales que l'on ne peut plus tirer de Saint-Domingue, percer le canal de Suez pour ouvrir une route commerciale entre l'Océan indien et la Méditerra-

1. Cette étude a été préparée et présentée au cours d'un détachement auprès du Laboratoire d'histoire des sciences et des techniques du CNRS (UPR 21).

2. Mais les beys mamelouks, réfugiés en haute Egypte, étaient revenus dès 1791. Voir Henry Laurens, *L'expédition d'Egypte, 1798-1801,* Paris, Armand Colin, 1989.

née, et détenir ainsi une position stratégique menaçant la puissance anglaise en Inde[3].

Sans faire remonter la modernisation de l'Egypte à l'expédition de Bonaparte, il est certain que celle-ci a été — par le choc des cultures sur le terrain comme par ses conséquences politiques à moyen terme — l'accident historique qui en a favorisé le développement dans les décennies suivantes[4]. La prise en compte du contexte et des enjeux intellectuels de l'expédition permet de mieux comprendre pourquoi Bonaparte emmène avec lui quelque 200 savants, ingénieurs et techniciens, organisés en une Commission des sciences et arts, et fonde, avec le mathématicien Monge, un Institut d'Egypte sur le modèle encyclopédique de ceux de Paris (établi moins de trois ans plus tôt) ou de Rome (établi par Monge quelques mois plus tôt seulement)[5].

Dans cette entreprise, la France pouvait avoir des atouts, notamment une expérience préalable de l'Empire ottoman. Sans compter l'alliance traditionnelle, depuis François Ier et Soliman, ni l'implantation commerciale dans les Echelles du Levant et en Egypte même — résultat des capitulations signées à l'époque — Bonaparte pouvait s'appuyer sur des hommes qui avaient récemment vécu et travaillé dans l'Empire ottoman, particulièrement dans les milieux consulaires ou dans les missions militaires envoyées tant par Louis XVI que par le gouvernement républicain[6]. Mais rares sont en fait les savants et ingénieurs de l'expédition d'Egypte qui ont une connaissance préalable de ce pays, et plus généralement de l'Empire ottoman ou de la culture musulmane. Moins nombreux encore, même chez les orientalistes, sont ceux qui en ont une

3. Voir aussi Patrice Bret, *L'Egypte au temps de l'expédition de Bonaparte (1798-1801),* Paris, Hachette Littératures, 1998 (coll. " La Vie quotidienne "), et les colloques du bicentenaire : P. Bret (éd.), *L'expédition d'Egypte, une entreprise des Lumières, 1798-1801,* Paris, Tec & Doc, 1999 ; *La campagne d'Egypte. Mythes et réalités, 1798-1801,* Paris, Ed. In Forma, 1998.

4. De plus, l'expédition et son chef d'oeuvre, la *Description de l'Egypte,* allaient devenir le modèle obligé des expéditions scientifiques ultérieures organisées dans le cadre d'opérations militaires, tant par la France (Morée, Algérie, Mexique) que, à une moindre échelle, par l'Egypte réformée de Muhammad 'Alî (Nubie, oasis de Sîwâ).

5. Afin de limiter la taille et le nombre de notes, et en l'absence d'autre indication, nous renvoyons, spécialement pour les références biographiques, au dictionnaire que nous préparons sur les aspects culturels de l'expédition d'Egypte.

6. Auguste Boppe, " La France et le militaire turc au XVIIIe siècle ", *Feuilles d'histoires du XVIIIe siècle* (1912), 386-402, 490-501 ; Frédéric Hitzel, *Le rôle des militaires français à Constantinople (1784-1798)*, Mémoire de maîtrise, Paris IV-Sorbonne, 1987 ; *idem,* " Une voie de pénétration des idées révolutionnaires : les militaires français à Istanbul ", *Varia Turcica, XIX* (1992), 87-94 ; *idem,* " Les écoles de mathématiques turques et l'aide française (1775-1798) ", dans D. Panzac (éd.), *Histoire économique et sociale de l'Empire ottoman et de la Turquie (1326-1960),* Paris, Peeters, 1995, 813-825 (Collection Turcica, 8) ; *idem,* " La formation des ingénieurs turcs ", communication présentée au *XXe Congrès International d'Histoire des Sciences, Liège, 20-26 Juillet 1997,* Symposium " La formation des ingénieurs en perspective : transfert institutionnel des modèles et réseaux de médiation (XIXe-XXe), organisateurs Gouzévitch, Grelon, Karvar ; *idem,* " La France et la modernisation de l'Empire Ottoman à la fin du XVIIIe siècle ", dans P. Bret (éd.), *L'expédition d'Egypte, une entreprise des Lumières, op. cit.* (sous presse) ; Mustafa Kaçar, *Osmanli Devleti'nde Mhendishanelerin Kurulusu ve Bilim ve Egitim Hayatindaki Degismeler* (L'établissement des écoles du Génie et les changements des systèmes éducatif et scientifique dans l'Empire ottoman), Thèse de doctorat, Institut des Sciences sociales, Université d'Istanbul, 1996.

expérience directe. Outre quelques marins qui ont navigué en Méditerranée orientale[7], quelques drogmans, négociants et consuls, c'est le cas d'une poignée d'ingénieurs qui ont déjà exercé leur activité sur le Bosphore ou dans les provinces arabes de l'Empire ottoman. Mon propos sera d'analyser brièvement dans quelle mesure cette expérience préalable a été réinvestie en Egypte.

Dans un premier temps, je retracerai rapidement la réalité de l'expérience de quelques membres de la Commission des sciences et arts. Je les suivrai ensuite en Egypte, pour constater et comprendre le décalage important entre l'oeuvre accomplie antérieurement en Turquie et leurs travaux en Egypte. Enfin, j'inverserai la problématique en cherchant comment l'expérience acquise durant les trois années de l'expédition a été elle-même réinvestie au XIXe siècle dans l'Empire ottoman, à l'heure de la question d'Orient, celle du démembrement et de la régénération de l'Empire ottoman se pose avec acuité.

L'EXPÉRIENCE DE L'EMPIRE OTTOMAN

En préparant l'expédition, Bonaparte fait appel à deux groupes qui ont été en contact avec l'Empire ottoman dans les années précédant l'expédition d'Egypte : l'un, dont l'homogénéité repose surtout sur une longue pratique du terrain, est celui des orientalistes ; l'autre, très hétérogène, se compose d'ingénieurs et architectes.

Les orientalistes de la Commission des sciences et arts

Parmi les orientalistes de la Commission, formée surtout de jeunes élèves de l'Ecole des langues orientales aux connaissances livresques, deux figures de proue représentent les deux principaux groupes d'orientalistes dont les services seront utilisés en Egypte : les vrais orientalistes, drogmans de profession, et ceux formés sur le terrain, généralement dans le négoce.

Au premier groupe appartient Jean-Michel Venture de Paradis (1739-1799), issu d'une vieille famille de drogmans et consuls par son père, et d'une famille grecque de Tripoli par sa mère. Agé de 59 ans en 1798, Venture totalise 35 années d'Orient, en poste à Constantinople, Saïda, au Caire (6 à 8 ans), à Tunis, en voyage en Syrie, ou en mission au Maroc et à Alger. Dans les cinq années précédant l'expédition, il a accompagné les ambassades de Verninac (1793) et d'Aubert-Dubayet (1796) à Constantinople, d'où il est rentré en compagnie de l'ambassadeur ottoman Morali Seyyid Alî efendi, pour occuper les chaires de turc et de tartare à l'Ecole des langues orientales, fondée en 1795 sur la base du plan de réorganisation de l'Ecole des Jeunes de Langues qu'il avait lui-même proposé. Outre sa longue expérience et ses compétences lin-

7. Par exemple le lieutenant de vaisseau provençal Blaise Rouden (1765-18 ?), qui avait récemment navigué jusqu'à Constantinople et dans les îles Ioniennes.

guistiques et diplomatiques certaines, Venture est aussi l'auteur d'un projet d'expédition en Egypte, rédigé en 1780, alors qu'il était en poste à Tunis[8].

De la même génération, le Marseillais Charles Magallon (1741-1820), représente le second groupe, dont les autres membres résident en Egypte. Il a séjourné plus de trente ans en Egypte, comme régisseur d'une maison de commerce textile marseillaise et premier député de la nation française dans cette province de l'Empire ottoman. Lié aux femmes des beys mamelouks, par l'intermédiaire de la sienne, il a acquis une influence certaine auprès de ces derniers. En mars 1789, alors qu'il est consul par intérim, Ismail pacha et le *shaykh al-balâd* Ismail bey lui réclament l'envoi d'une mission militaire française semblable à celles qui ont été envoyées à Constantinople, avec génie, artillerie, fonderie et constructions navales. La mort du second (avril 1791) et le retour d'Ibrâhîm bey et de Murâd bey met un terme à cette période faste, car les maladresses du consul Mure, au lendemain de l'intervention de Hasan pacha, ont malheureusement ruiné le climat de confiance antérieur. Rentré à Paris (été 1791) pour présenter les doléances des marchands français, Magallon repart au printemps 1793, comme consul de France en Egypte, en poste à Alexandrie, et ne cesse dès lors de prôner une intervention française auprès de Verninac, envoyé de la République à Constantinople (juin 1795), et de Colchen, commissaire des relations extérieures (octobre). Au même moment, il accompagne au Caire l'envoyé du Comité de salut public Dubois-Thainville, avec le négociant Baudeuf (octobre 1795). Ils y sont reçus avec tous les honneurs par le pacha et les deux beys mamelouks, en janvier 1796, mais Dubois-Thainville repart quatre mois plus tard sans être parvenu à faire rembourser aux Français les emprunts forcés prélevés par les mamelouks. Magallon s'embarque à nouveau pour la France en juillet 1797. Le rapport qu'il remet à son retour sera décisif pour convaincre le gouvernement d'accepter le projet d'expédition élaboré par Talleyrand et Bonaparte.

Les ingénieurs

Au moment d'aborder le groupe des ingénieurs et architectes qui ont participé dans les années 1780 et 1790 à des missions françaises dans l'Empire ottoman, une question reste sans réponse : pourquoi ne pas avoir engagé systématiquement tous les officiers ayant appartenu à ces missions ? Lafitte-Clavé, devenu général, est mort en 1793, mais certains de ses collaborateurs principaux sont toujours disponibles, tels l'ingénieur militaire Joseph-Gabriel Monnier de Courtois (1745-1818) — d'ailleurs retourné en Turquie de 1794 à 1797 — ou Charles-François Frérot d'Abancourt (1758-1801), ingénieur-géographe des Affaires Etrangères, qui a passé deux ans et demi en Turquie (1785-

8. Ezzedine Guellouz, " Analyse idéologique d'un projet d'expédition d'Egypte : le projet de Venture de Paradis, orientaliste philosophe (1780) ", *Les Cahiers de Tunisie*, n° 81-82, 21 (1973), 123-153 (projet : 138-151).

1788). Il ne semble pas, par exemple, que ce dernier, devenu directeur de la section topographique du Dépôt général de la Guerre, ait jamais été sollicité, alors que cette institution envoya la brigade topographique d'Italie.

Toutefois, trois ingénieurs, deux architectes et un officier de l'état-major de Bonaparte, ont déjà séjourné dans l'Empire ottoman, l'un sous Louis XVI, roi de 1774 à 1792, et 'Abdül-Hamîd Ier, sultan de 1774 à 1789, les autres sous la République et Selîm III sultan de 1789 à 1808.

L'ingénieur constructeur de vaisseaux Jean-Jacques-Sébastien Le Roy (1747-1825) a passé plus de quatre ans en mission en Turquie, de 1784 à 1788, échappant à la peste, qui tue son domestique et son adjoint Du Rest. Selon le ministre des Affaires étrangères, le comte de Montmorin, " a le mérite d'exciter la confiance des Turcs, et de leur faire faire plus qu'on ne devoit attendre de leur ignorance et de leurs préjugés "[9]. Après un voyage dans les forêts d'Anatolie, il établit une salle des gabarits, où il forme quelques élèves et fait exécuter les gabarits des vaisseaux de tous rangs et des frégates de 24, 18 et 12, dont il a donné les plans et les instructions. A Constantinople même, il construit un vaisseau de 74, lancé en 1787 — et refond aux normes françaises un vaisseau turc en chantier —, quatre corvettes, une galiote et une prame et quelque quatre-vingts bombardes et canonnières. Parallèlement, il dirige les travaux de radoubs et d'armements. Son action est reprise en 1793 par Jacques-Balthasard Brun de Sainte-Catherine (1759-?)[10].

Les officiers du génie membres de missions militaires de la Révolution qui suivent Bonaparte en Egypte n'ont pas un tel succès dans leurs travaux en Turquie, quoiqu'il s'agisse de sujets brillants que leur carrière ultérieure mènera jusqu'au grade de général. Ingénieur des ponts et chaussées, devenu capitaine du génie (plus tard, général), Joseph-Félix Lazowski (1759-1812), d'une famille polonaise tenant un office à la cour du roi Stanislas de Lorraine, envoyé en mission à Constantinople à la place de Mazurier, passe plus de deux ans et demi sur le Bosphore, qu'il quitte en juillet 1797. Le rapport sur l'état de la Turquie qu'il rédige en décembre prêche aussi en faveur de l'intervention[11]. Quant au séjour d'Henri-Gatien Bertrand (1773-1844), également capitaine du génie, il se limite à quelques mois, et sa connaissance du pays reste donc superficielle : arrivé avec l'ambassade du général Aubert-Dubayet, qui arrive à Constantinople à la fin de l'année 1796, il rentre dès le mois de mai 1797.

Deux architectes, Jean-Baptiste Lepère (1761-1844) et Jean-Constantin Protain (1769-1837), appartiennent à l'importante mission Pampelonne, qui

9. Lettre au maréchal de Castries, 1er mars 1787.

10. Stanford J. Shaw, " Selim III and the Ottoman Navy ", *Turcica,* I (1969), 212-241 ; F. Hitzel, " La France et la modernisation de l'Empire Ottoman à la fin du XVIIIe siècle ", *op. cit.*

11. " Observations sur l'état présent de la Turquie et sur l'alliance de cette puissance avec la République française " (Service historique de l'armée de terre, Archives du génie, Vincennes (ci-après : SHAT), Art. 14, Turquie, carton 2, f° 14). Voir Adam M. Skalkowski, *Les Polonais en Egypte, 1798-1801,* Paris, Grasset, 1901.

accompagne aussi cette ambassade[12]. Ils sont au nombre des vingt-cinq artistes — le tiers de cette mission seulement — qui sont effectivement employés par les Turcs. Député au Conseil des cinq-cents et ancien directeur d'une fonderie de canons à Valence, Pampelonne devait notamment établir à Constantinople une nouvelle fonderie, dont Lepère dresse les plans, achevés à l'automne 1797. Le projet n'ayant finalement pas de suite, Lepère rentre en visitant la Thrace, la Bosnie, la Dalmatie et l'Italie du Nord, après avoir profité de son séjour pour dessiner les monuments et les établissements militaires stambouliotes[13].

Un autre officier d'état-major pourrait être associé à cette catégorie des ingénieurs et architectes, tout en ayant des liens avec les orientalistes : passé en France en janvier 1793, le jeune officier révolutionnaire polonais Joseph Sulkowski (1770-1798) s'est aussitôt lié avec l'un des ses compatriotes, Maleszéwski, gendre de Venture de Paradis, et il aurait lui-même épousé la seconde fille de ce dernier. Envoyé cinq mois plus tard en Turquie à la suite de l'ambassade de Sémonville, il reçoit finalement l'ordre de se rendre en Syrie où de nouvelles instructions pour une mission aux Indes lui seront adressées. Sulkowski commence à apprendre le turc et part pour Constantinople et Alep. Là, il attend en vain les nouveaux ordres, apprenant l'arabe et se familiarisant avec la vie orientale, avant de regagner Constantinople, qu'il quitte en septembre 1794 pour la Pologne. Passé en Italie à sa demande en 1796, il devient aide-de-camp de Bonaparte, et le suit naturellement en Egypte. Polyglotte — outre sa langue maternelle et les langues de l'Empire ottoman, il parle aussi français, russe, allemand, anglais, italien et espagnol —, d'esprit aventureux, vaillant et intelligent, Sulkowski était particulièrement indiqué pour le projet égyptien. Il participa au rassemblement de la documentation, imprimés et cartes, pour l'expédition[14].

12. Voir F. Hitzel, " La France et la modernisation de l'Empire Ottoman à la fin du XVIIIe siècle ", *op. cit.* et " Une voie de pénétration des idées révolutionnaires : les militaires français à Istanbul ", *op. cit.*. Sur cette mission en général, voir F. Clément-Simon, *Le premier ambassadeur de la République francaise à Constantinople, le général Aubert Duhayet,* Paris, Impr. de la Renaissance latine, 1904 ; Fazi du Bayet, *Les généraux Aubert Dubayet, Carra Saint Cyr et Charpentier,* Paris, 1902 ; Auguste Dry, *Soldats ambassadeurs sous le Directoire,* Paris, Plon, 1906, t. I ; P. Odinot, " Contribution à l'histoire de l'artillerie française, une mission d'artillerie française à Constantinople en 1797 ", *Revue d'artillerie,* n° 97 (mai 1926), 492-505. Un artiste mécanicien de la Commission des sciences et arts, François-Sébastien Aimé (1762-1843), alors employé à l'Ecole polytechnique aurait dû partir également avec cette mission : commissionné à cet effet par la Marine, en août 1796, il fut finalement appelé à d'autres fonctions. *Cf.* P. Bret, " Les oubliés de Polytechnique en Egypte : les " artistes " mécaniciens de la Commission des Sciences et des Arts ", dans *Scientifiques et sociétés pendant la Révolution et l'Empire*, Paris, CTHS, 1990, 497-514.

13. *Cf.* W. Müller-Wiener, " Jean-Baptiste Lepère (1761-1844) in Istanbul. Zum Fruhwerk eines Baumeisters des Klassizismus ", *Festschrift Gerhard Bott* (Nurnberg, 1987), 103-113 ; Uwe Westfehling, " Les dessins de Lepère au Musée Wallraf-Richartz de Cologne ", dans *L'Egypte, une description* (cat. expo. Musée Fesch, Ajaccio : Ville d'Ajaccio, 1988), 175-181.

14. *Ibid.* Voir aussi Hortensius Corbeau de Saint-Albin, *J. Sulkowski, chef de brigade, aide-de-camp du général Bonaparte et Membre de l'Institut d'Egypte. Mémoires historiques, politiques et militaires sur les révolutions de Pologne, 1792-1794, la campagne d'Italie, 1796-1797, et les campagnes d'Egypte, 1798-1799,* Paris, 1832, 2 vols ; M. Reinhard, *Avec Bonaparte en Italie, d'après les lettres inédites de son aide de camp Joseph Sulkowski,* Paris, Hachette, 1946.

D'une expérience à l'autre

Presque tous ces hommes d'expérience furent membres de l'Institut d'Egypte créé par Bonaparte, les uns dès l'origine (Le Roy en section de mathématiques, Sulkowski en économie politique, et Venture en Littérature et arts), les autres lors d'élections complémentaires, dans les sections de physique (Beauchamp, le 7 septembre 1798) et de Littérature et arts (Lepère, le 1er décembre 1798 ; Protain, le 21 janvier 1800).

Trois types de missions principales devaient être confiées à ces hommes d'expérience[15]. Deux relèvent directement de la profession de certains : par métier, les orientalistes assurent l'interprétariat et la traduction, et les ingénieurs militaires effectuent les reconnaissances du territoire. Le troisième type est plus spécifique à l'expédition : l'expérience des orientalistes et leur connaissance des us et coutumes du Moyen-Orient pouvait aussi les rendre particulièrement utiles dans les négociations et les contacts diplomatiques, dont furent également chargés quelques " Francs " vivant au Caire ou arrivés en Egypte par leurs propres moyens. En revanche, si l'on pouvait s'attendre à voir les ingénieurs et architectes participer à la régénération aux côtés des Egyptiens, comme ils l'avaient fait aux côtés des Turcs, notamment par la formation aux techniques européennes, il n'en fut rien.

Les occupations professionnelles ordinaires des ingénieurs et architectes

L'examen des travaux de chacun des personnages déjà mentionnés est à cet égard probant. Par exemple, Le Roy, dont l'oeuvre à Constantinople avait été la plus importante et la plus durable, pour la modernisation de la marine turque, joue dans l'expédition un rôle non négligeable, mais tel qu'il aurait pu le faire n'importe où ailleurs : dans ses nouvelles fonctions d'ordonnateur en chef de la marine (et préfet maritime, en 1800), il a préparé la flotte, créé un établissement maritime à Alexandrie, supervisé les quelques constructions et réparations navales à Suez, Damiette et Bûlâq, outre quelques autres fonctions

15. L'expérience de l'Empire ottoman n'est pas seule recherchée. Compte aussi celle des Tropiques, une expérience coloniale acquise à Saint-Domingue. L'architecte Lepère la possède également, pour y avoir passé quatre ans (1787-1790), construisant des habitations. Mais c'est surtout le cas du jardinier botaniste Nectoux, qui remplace le professeur de culture du Muséum, Thouin, initialement porté sur la liste des savants, et qui a passé neuf ans en Guyane et à Saint-Domingue. Sur ce dernier, voir P. Bret, " Le réseau des jardins coloniaux : Hypolite Nectoux (1759-1836) et la botanique tropicale de la mer des Caraïbes aux bords du Nil ", dans Y. Laissus (éd.), *Les naturalistes français en Amérique du Sud, XVIe-XIXe siècles,* Paris, CTHS, 1995, 185-216 ; " La plantation idéale des Lumières : Nature, esthétique et équilibre dans la caféière du jardinier-botaniste Nectoux ", dans J.-Cl. Hocquet (éd.), *Colloque du 123e Congrès national des sociétés historiques et scientifiques, Fort-de-France, avril 1998, " Le sucre et l'économie de plantation "*, Paris, CTHS (sous presse) ; " Des " Indes " en Méditerranée ? L'utopie tropicale d'un jardinier des Lumières et la maîtrise agricole du territoire ", dans M.-N. Bourguet et C. Bonneuil, *Revue française d'histoire d'outre-mer,* 86, n° 322-323 (" Botanique et colonies ") (1999), sous presse.

administratives comme la Commission des hôpitaux et l'administration sanitaire.

La connaissance de l'architecture turque qu'avaient acquise les architectes Lepère et Protain ne fut probablement pas primordiale dans les attributions qu'ils reçurent. Lepère, chargé des bâtiments militaires de la province du Caire, eut surtout à convertir des maisons en casernes, mais érigea aussi les moulins à vent de la capitale, dont l'ingénieur Conté réalisa la mécanique, avant d'être nommé, en janvier 1801, directeur des grandes fouilles de Giza. Protain, qui l'assista dans ses premières fonctions, fut membre de la Commission des renseignements sur l'état moderne de l'Egypte, chargé des monuments. Responsable des aménagements que Kléber fit faire au palais du quartier-général, il fut blessé en tentant de s'interposer lorsque celui-ci fut assassiné au cours d'une visite des travaux. Menou le nomma architecte national et conservateur des " objets appartenant à la République dont le g[énér]al en chef aura la jouissance ".

Quant à Lazowski, Bertrand et Sulkowski, ils ne remplirent que les missions ordinaires de leur profession. Ce dernier fut toutefois nommé, lors de la deuxième séance de l'Institut, membre d'une commission chargée de préparer un vocabulaire arabe, avec Desgenettes et Tallien, et sans aucun orientaliste !

Les orientalistes de retour sur le terrain

Quant à ceux-ci, leur tâche était toute trouvée, et il est inutile de faire la liste de leurs interventions. Deux points méritent cependant d'être soulignés.

D'une part, l'importance de Venture de Paradis dépassa largement sa fonction d'interprète. Le général du génie Caffarelli du Falga, responsable de la Commission des sciences et arts, déclarait " qu'il ne connaissait pas d'homme plus utile à l'armée d'Orient que Venture et Conté ", et l'orientaliste et directeur de l'Imprimerie nationale du Caire Marcel a écrit de lui : " Premier interprète du général en chef, il fut plutôt son premier ministre pour tout ce qui concernait le pays et les populations de l'Orient. Chéri de tous les Français qui l'approchaient, il avait su se créer une grande influence sur tous les musulmans, juifs et chrétiens de l'Egypte et de la Syrie. "[16] Le secrétaire de Bonaparte, Bourrienne, a noté combien était forte l'influence que Venture avait sur le général en chef en matière de relation avec les musulmans, depuis la déclaration islamophile rédigée avant le débarquement jusqu'aux diverses proclamations destinées aux Egyptiens[17]. La mort de cet homme d'expérience a été une perte considérable, que ne pouvaient compenser ni les talents des jeunes inter-

16. Cité par Henri Dehérain, " L'Egypte turque. Pachas et mameluks du XVIe au XVIIIe siècle. L'expédition du général Bonaparte ", dans G. Hanotaux, *Histoire de la nation égyptienne,* Paris, Plon-Nourrit, 1931, t. 5, 309.

17. Bourrienne, *Mémoires de M. de Bourrienne, ministre d'Etat, sur Napoléon, le Directoire, le Consulat, l'Empire et la Restauration,* Paris, Ladvocat, 1829, II, 158-159. Voir aussi H. Laurens, *op. cit.*, 1989.

prètes Jean-Joseph Marcel (1776-1854), Amédée Jaubert (1779-1847), Nicolas-Henry Belleteste (1778-1808), Jacques-Denis Delaporte (1777-1861) et Louis-Rémy Raige (1777-1810), ni le recours aux drogmans des consulats de la région, comme Damien Bracevich (177?-1830) à Alexandrie ou Jean-Baptiste Santi L'Homaca à Jaffa[18].

Le rôle de Magallon est moins important, quoiqu'il soit également chargé de négociations ponctuelles. Mais il est, en quelque sorte, l'orientaliste de la Commission administrative provisoire, qu'il forme avec Monge et Berthollet dès juillet 1798, bientôt le premier des cinq administrateurs de l'Enregistrement et des Domaines nationaux, et *wâlî* de Girga en janvier 1799.

Une mission diplomatique particulièrement délicate fut confiée à un autre orientaliste, l'astronome Joseph Beauchamp (1752-1801), correspondant de l'Académie royale des sciences de Paris puis de l'Institut national, qui a passé près de neuf ans à faire des observations en Mésopotamie, de 1781 à 1789. Nommé consul à Mascate, il part à l'automne 1796 pour rejoindre son poste, voyageant sur les bords de la mer Noire entre deux séjours à Constantinople, qu'il quitte en novembre 1797 pour la Syrie lorsque la flotte française atteint Alexandrie. Se détournant alors de sa route, il arrive au Caire après l'installation des Français et participe aux travaux de l'Institut d'Egypte, notamment pour la mettre en place un observatoire avec l'astronome Nouet. Au moment de pénétrer en Syrie, en février 1799, Bonaparte le charge d'une mission auprès du grand vizir à Constantinople. Capturé par les Anglais, il reste emprisonné à Constantinople jusqu'en septembre 1801 : libéré par la capitulation de l'armée d'Orient, il meurt bientôt sur le chemin du retour.

Le renfort des " Francs " d'Egypte

L'expérience du terrain de quatre autres négociants français établis en Egypte est également utilisée[19]. François Baudeuf, principal négociant français du Caire, faisait office de consul officieux au Caire depuis le transfert du consulat à Alexandrie. Il avait déjà joué les bons offices lors de la mission de Dubois-Thainville en 1795-1796 et juste après la bataille des Pyramides. Il fut l'interprète du général Dupuy, commandant la place du Caire, membre du deuxième *Dîwân* en décembre 1798 et secrétaire de la Commission des renseignements sur l'état moderne de l'Egypte onze mois plus tard. Le Lorrain Louis-Elie Caffe, négociant du Caire et de Rosette, appartint même aux deux premiers *Dîwân* (juillet et décembre 1798). Il fut nommé par Commission des subsistances pour la vente des grains et participa à la définition du rapport entre l'ardeb et les poids français. Leur confrère Naydorff, fils d'un négociant

18. Voir Gabriel Guémard, " Les orientalistes de l'Armée d'Orient ", *Revue de l'Histoire des Colonies françaises,* 21/1 (1928), 129-150.

19. De même que celle du neveu de Magallon, Jean-André Magallon, vice-consul à Alexandrie.

en grains, fut secrétaire interprète de l'intendant général copte Jârkis al-Jawharî en août 1798, puis interprète de la Commission des renseignements, à la fin de l'année suivante et fermier du gouvernement pour les grains de la Haute-Egypte. Enfin, à Rosette, le Marseillais Joseph Varsy, agent consulaire français à Rosette, a fait préparer l'arrivée des Français et reçoit plusieurs membres de la Commission des sciences et arts dans sa maison. Dugua le nomme membre de la Commission administrative provisoire de Rosette, avec un négociant copte et trois notables musulmans de la ville, puis du *Dîwân* local.

A ces négociants-interprètes, il faudrait associer les religieux européens ou syriens qui ont pareillement servi d'interprètes. Le premier d'entre eux est le prêtre syrien de rite grec catholique (melkite) Rapha'l de Monachis (Rapha'l Antoun Zakhûr Râhib, 1759-1831), secrétaire du père général de l'ordre des Basiliens, qui sera le seul membre oriental de l'Institut d'Egypte, le traducteur de nombreux textes émis par le pouvoir et par les savants (comme l'Avis sur la petite vérole du médecin en chef Desgenettes), le premier drogman du *Dîwân* sous Menou, et le principal lien direct entre la pseudo-régénération voulue par les Français et la modernisation effectuée par Muhammad 'Alî[20]. L'emploi d'autres religieux est plus ponctuel, comme celui du père Antonio par le général Belliard à Qûsayr.

Deux autres " Francs " établis jouèrent un rôle plus ou moins grand dans les institutions mises en place par les Français. Carlo de Rosetti (1736-1820) est le premier des Francs, un personnage mieux établi encore. Négociant vénitien établi très jeune en Egypte, il entre au consulat d'Autriche et de Toscane en 1760 et devient consul général de ces deux pays en 1784. Il a d'abord la confiance d''Alî bey, qui gouverne de 1763 à 1771 et auquel il sert de conseiller pour le commerce avec l'Inde, et sous lequel les Vénitiens — les plus nombreux après les Français — obtiennent la faveur des autorités. Il devient le confident de Murâd bey, qui lui concède deux grands monopoles affermés, le séné et le natron. Il est alors un homme incontournable en Egypte pour les puissances européennes et les voyageurs, et le sera pour l'occupant. Source de renseignements, pour les autorités comme pour les membres de la Commission des sciences et arts — notamment Nectoux —, il ne fut pas membre du *Dîwân,* mais participa à deux institutions, le bureau de santé et de salubrité (30 septembre 1798) et la Commission des renseignements sur l'état moderne de l'Egypte, pour les articles Gouvernement et histoire (avec Fourier) et Commerce et industrie (avec Livron, puis Girard et Conté) treize mois plus tard (novembre 1799). Le second, Enrico di Wolmar (1749-ap. 1827), médecin italien d'origine balte, a voyagé dans les Etats barbaresques et l'Empire ottoman — en Turquie et en Grèce — avant d'arriver en Egypte, où il réside depuis

20. Charles Alexandre Bachatly, " Un membre oriental du premier Institut d'Egypte, Don Rapha'l (1759-1831) ", *Bulletin de l'Institut d'Egypte,* 17 (1935), 237-260.

1788, et est devenu le médecin particulier de Murâd bey. Il est nommé membre du deuxième *Dîwân* en décembre 1798.

Deux autres médecins sont en Egypte lors de l'arrivée des Français. L'un d'eux, Réal, semble déjà exercer la médecine à Alexandrie ; attaché aux lazarets, il semble être à la fois médecin et interprète. L'autre, Louis Frank (1761-1825), d'une famille de médecins célèbres, voyage en Egypte, étudiant la pharmacopée locale, lorsque Desgenettes le fait intégrer comme médecin militaire et membre du Bureau de santé et de salubrité dès septembre 1798. Il donne un mémoire sur le grand Caire dans le cadre de la Topographie physique et médicale de l'Egypte, et présente quatre mémoires à l'Institut d'Egypte.

L'expérience de l'expédition

En définitive, l'expérience acquise durant l'expédition a été plus déterminante pour la carrière de certains. Si Magallon poursuit sa carrière en quittant l'Egypte, pour devenir consul général à Salonique en 1802, plusieurs membres de l'expédition font leurs premières armes sur le terrain. C'est d'abord le cas pour les interprètes Jaubert et Delaporte, appelés à une longue carrière qui les mènera à dépasser l'interprétariat. Chargé de missions diplomatiques en Turquie et en Perse au début de l'Empire, Jaubert est membre de l'Académie des Inscriptions et belles lettres, professeur au collège de France, directeur de l'Ecole des Langues orientales et pair de France sous Louis-Philippe ; chancelier-interprète et consul à Tripoli sous l'Empire et à Tanger sous la Restauration, Delaporte est interprète en chef de l'Armée d'Afrique et directeur des Affaires arabes à Alger, puis consul à Mogador sous Louis-Philippe. Plusieurs autres membres ou proches de la Commission des sciences et arts se découvrent alors une vocation nouvelle, à commencer par les chirurgiens Hugues Pouqueville (1770-1838) et Julien Bessières (1777-1840), capturés et retenus prisonniers avant de retourner en mission diplomatique en 1806 à Janina, où Frank est médecin d''Ali pacha. L'élève astronome Jérôme-Isaac Méchain (1778-185?), le chimiste et ingénieur Joseph-Angélique-Sébastien Regnault (1776-1827), le mathématicien Louis-Alexandre Olivier de Corancez (1770-1832) entrent aussi dans la carrière consulaire, qui les conduira tous au Moyen-Orient ou au Maghreb : le premier en Moldavie, aux Dardanelles, à Chypre et à Tanger, le deuxième en Crète, à Chypre et en Syrie, le troisième en Syrie.

La carrière diplomatique du général Andréossy tient sans doute à d'autres facteurs — il est titulaire de l'ambassade de Londres durant la paix d'Amiens — mais sa connaissance de l'Egypte a pu n'être pas étrangère au choix que fit Napoléon en l'envoyant comme ambassadeur à Constantinople en 1812. Quant au chevalier de Malte Théodore-Jules Lascaris (1765-1817), sa participation à l'expédition, a été déterminante dans sa fascination certaine pour l'Orient. Tour à tour architecte de l'administration de l'Enregistrement et des Domaines nationaux, directeur des adjudications et droits sur les biens meubles des successions (Beyt-el-Mal), et directeur des droits affermés, il apprend l'arabe et se lie au *mu'allim* Ya'qûb, jusqu'à épouser sa cause. De retour au Moyen-Orient en même temps qu'Andréossy, il explore secrètement les possibilités d'une expédition vers les Indes avec l'accord de la Porte et passe le restant de ses jours entre la Syrie — où il fréquente Lady Stanhope — l'Asie mineure ou les Cyclades, avant de retourner en Egypte, en 1816. Par l'intermédiaire du consul Drovetti, il devient précepteur de français du prince Ismaïl, second fils de Muhammad 'Alî, en butte à l'hostilité du ministre Boghos Youssouf, et meurt au Caire.

On pourrait encore citer le pharmacien militaire Claude Royer, qui reste en Egypte après l'évacuation, et meurt au Caire, ou le général Boyer, qui y revient à la tête d'une mission officielle sous la Restauration. Mais, finalement, et bien qu'il ne retourne jamais en Egypte, l'itinéraire personnel le plus marqué par le passage en Egypte est celui de l'ingénieur géographe Edme-François Jomard (1777-1862), qui dirige la publication de la *Description de l'Egypte* à partir de 1808, crée la Mission scolaire égyptienne à Paris, pour accueillir des étudiants égyptiens à partir de 1826, et qui anime non seulement le réseau des " Egyptiens " de la Commission des sciences et arts, mais un véritable lobby égyptien en France pour appuyer la politique de modernisation de Muhammad 'Alî.

Dans un contexte purement français, l'oeuvre scientifique de l'expédition de Bonaparte s'inscrit en continuité, quoiqu'à une échelle sans commune mesure, avec les grandes expéditions scientifiques maritimes de la période — La Pérouse, d'Entrecasteaux, Baudin — et avec les petites Commissions des sciences et arts associées aux armées de la Révolution, en Belgique et en Italie. Dans le contexte géopolitique des relations avec la Porte, au contraire, elle s'inscrit en rupture avec les missions militaires françaises envoyées à Constantinople sous les sultans Mustafâ III, 'Abdül-Hamîd Ier et Selîm III, illustrées surtout par la création, en 1784, de l'école du génie *(Mühendishâne-i-Hümâyûn)* : en 1798, la France déplace son action du centre politique de l'Empire vers une périphérie convoitée sur la route des Indes et, de fait, malgré le discours loyaliste, elle joue pour la première fois la carte du démembrement de l'Empire ottoman. Dans ces conditions, et malgré les velléités de régénération, l'expérience acquise lors des missions antérieures était sans grand fondement, sinon pour les problèmes linguistiques.

Aussi, puisque, à part les orientalistes, les hommes ayant une expérience préalable de l'Empire ottoman sont des hommes de la technique, je propose, en guise de conclusion, une réflexion autour de trois objets techniques emblématiques de l'expédition, qui la résument : le moulin à vent, le moulin à poudre, le moulin à plâtre.

Le premier, issu du Moyen-Orient et particulièrement abondant dans les îles de l'Empire ottoman, est, dès l'arrivée en Egypte, " le premier bienfait que les artistes d'Europe se proposent d'accorder à l'Egypte [...] où tant de bras, utiles aux arts ou à la culture, sont employés à élever l'eau du fleuve, à moudre lentement et mal un peu de grain "[21]. Ne passant sur les bords du Nil que par le biais des Français, il s'inscrit désormais dans le paysage comme la marque visible de la puissance et de l'efficacité technique européenne. Jabartî et Nicolas Turk — qui, comme les Cairotes, dédaignent la montgolfière — l'admirent, et l'imaginaire collectif égyptien en a conservé le souvenir, puisque les ruines des moulins qui dominent encore celles de la Fustât des Califes sont attribuées à Bonaparte, alors qu'il s'agit de constructions faites à la fin du règne de Muhammad 'Alî. Ainsi, le moulin à vent symbolise-t-il l'écart de la place accordée à la technique en France et en Egypte, et le transfert technique de la première vers la seconde qui se poursuivra au XIXe siècle. Construit dans les ateliers de mécanique du Caire, où Conté emploie une main-d'oeuvre locale et forme de jeunes apprentis égyptiens, il est l'emblème même de la régénération, dont il marque aussi les limites.

Le moulin à poudre, construit par les Français sur l'île de Roda rappelle d'abord que la présence française est une occupation militaire. Le remploi d'une mosquée ruinée (Qaïd bey) symbolise la laïcité ou l'athéisme de la République qui honore officiellement par ailleurs le Coran, le Prophète et les *shaykh's* d'al-Azhar. Le choix d'une île sur le Nil, point névralgique entre deux ponts de bateaux protégés par des installations militaires, et où fut projetée la construction d'une ville française, symbolise le repli colonial, la création d'un espace propre aux colons. La poudrerie de Roda marque aussi l'ambiguïté des relations franco-égyptiennes, puisqu'elle est aussi un des lieux de transfert technique, avec une main-d'oeuvre indigène, compromis entre les techniques françaises les plus récentes — les procédés révolutionnaires créés en 1794 — et des gestes techniques égyptiens traditionnels. Mais le transfert réel attendra la création par Pascal Coste, à la demande de Muhammad 'Alî et sur cette même île de Roda, d'une nouvelle poudrerie conçue et construite en 1820, selon la technique française classique[22].

Enfin, le moulin à plâtre, vient inverser le courant du transfert de techniques, puisqu'il s'agit, en l'occurrence, d'une technique égyptienne que les ingénieurs français vont rapporter en France : à Paris, le plâtre était encore broyé à bras d'homme, à la masse. Certes des machines existaient déjà dans les provinces méridionales et les pays limitrophes de la France. Mais la technique égyptienne qui, par un double mouvement, broyait et pilait le plâtre en une seule opération, se révéla supérieure, et Conté la fit adopter par la Société d'encouragement pour l'industrie nationale au retour de l'expédition.

21. Jean-François Detroye, " Journal ", 21 fructidor an 6/7 septembre 1798 (SHAT, MR 526-527) ; passage reproduit dans Clément de La Jonquière, *L'expédition d'Egypte, 1798-1801,* Paris, Lavauzelle, 1899-1905, t. III, 15.

22. P. Bret, " French Gunpowder Mills in Cairo during Bonaparte's Expedition (1798-1801) and Muhammad 'Alî's reign : Adaptation and Transfer of Technology ", dans Brenda J. Buchanan (éd.), *Actes du 25e congrès de l'ICOHTEC*, Lisbonne, 1998 (à paraître).

Pour être le plus symbolique, ce dernier exemple n'est pas unique[23]. Il montre que les voies du transfert au XIXe siècle sont plus complexes qu'un flux à sens unique en provenance d'une Europe technicienne. Il illustre un autre apport de l'expérience acquise durant l'expédition d'Egypte.

23. P. Bret, " La Méditerranée médiatrice des techniques : regards et transferts croisés durant l'expédition d'Egypte (1798-1801) ", dans Vassilis Panayatopoulos (éd.), *Les expéditions scientifiques françaises en Méditerranée* (Colloque de l'Ecole française d'Athènes et de l'Institut d'Etudes néo-helléniques, Athènes, juin 1995), Athènes, Institut d'Etudes neo-helléniques (à paraître).

The Introduction and Reception of Modern Science and Technology in " Ottoman " Egypt in the Nineteenth Century

A.H. Helmy MOHAMMAD

The Islamic World has actively sought to acquire scientific knowledge from the outside world in two major historical movements. The first was the great era of the selective translation of the basic scientific works of Classical Antiquity into Arabic, between the eighth and tenth centuries A.D. The second has been the influx of Western science during the last two centuries.

The first era is regarded as " one of the most remarkable instances of cultural transmission in human history " (Nasr, 1976, 12), and has been cherished by historians of science from both the East and West, who have also attended to the glorious achievements of Islamic science in its golden age and noted the translation of Arabic works into Latin. However, Professor Ekmeleddin Ihsanoğlu has rightly called our attention to the fact that the second era of translation and transfer of scientific knowledge to the Muslim World had been unjustly neglected and should be properly studied and analyzed (Ihsanoğlu, 1988A,[1] 1992). Professor Ihsanoğlu himself leads an active research program devoted to the study this era, and has initiated a number of symposia on it. These symposia have been organized by Research Center for Islamic History, Art, and Culture (IRCICA), with the later collaboration of the Turkish Society for the History of Science (TBTK) and other international partners.

My contribution to the present symposium is an attempt to shed light on some aspects of this subject with particular regard to " Ottoman " Egypt. In the eighteenth and nineteenth centuries, the relationship between the Ottoman Sultan and the direct local rulers of Egypt (the Mamluks and later Mohammed Ali's line) was in constant flux — though the fact that Egypt had a special semi-independent status within the Ottoman Empire meant that usually this

1. The letter A after the date of a reference indicates that the reference is to be seen in the list of references in Arabic.

relationship was not very strong. Nevertheless, cultural exchange was taking place, as will become apparent in the following discussion. This relationship of cultural and intellectual exchange was effectively severed by the British invasion of Egypt in 1882.

European influence had long been a strong and assertive presence in the field of intellectual exchange. Francis Robinson (1996) distinguishes two ages in the modern history of the Islamic World : the " Age of European Expansion " (1500-1800), and the " Age of Western Domination " thereafter. At the line between these two " ages " is the brief French invasion of Egypt (1798-1801). In this rather abrupt confrontation between East and West, modern Western science was displayed in Egypt with a great flourish. Bonaparte founded L'Institut d'Egypte (which still to this day exists and is active) — but the French took their press home with them when they departed (Aida Nosseir, 1994A). The impact on Egyptians of the great French work, *Description de l'Egypte* (1802) and the discovery of the Rashīd (Rosetta) Stone became manifest only later. At the time these were regarded solely as French achievements. However, direct contact between the French and Egyptian scholars did occasionally take place, as vividly portrayed by al-Jabartī in his well-known historical diary of 5th December, 1798 (al-Baqly, 1958A, 284-286). In this connection we should not forget Professor Ihsanoğlu's reminder of early translations from European languages to Arabic, Turkish, and Persian of works in the fields of geography, medicine, astronomy, and military matters, including translations dating to the fifteenth, sixteenth and seventeenth centuries (Ihsanoğlu, 1988A).

As I am not a professional historian, I decided to trace and assess the transfer of science and technology to Egypt during the Ottoman period along the following themes : technical and higher education, translation and book publication, introduction of modern technology, and the popularization of science through journalism. I originally intended to confine my analysis to the introduction and reception of biological concepts, while some other colleagues from Egypt dealt with physical, technological, and industrial concepts. However, as that arrangement could not be realized I am going to touch occasionally on those matters as well.

The ambitious Mohammad Ali Pasha, who ruled Egypt between 1805 and 1849, aimed to lay the basic foundations of a modern Egyptian state. In this endeavour he was strongly supported by a competent army and a dependable naval fleet. In 1827, responding to advice from Antoine Clot Bey, he founded a Medical School attached to the Army Hospital, which he followed up on in 1832 with a School of Pharmacy — complete with a garden for the cultivation of medicinal plants. In 1937 the two schools were transferred to the site of the present Faculty of Medicine and attached University Hospital at Kasr al-Aini. A year later (1838) a School for Obstetrics (or Midwifery) with an annexed hospital for women was established.

The teaching staff in medicine and pharmacy instructed students in physics, chemistry, botany, anatomy, and physiology. Students were selected from among the most distinguished pupils at al-Azhar, and the course of study in medicine was five years. The first cohort of graduates in medicine numbered 100, while the number of students in the School of Pharmacy was 50 in 1837 (Ibrahīm Badran *et al.*, 1995A). Mohammad Ali also established a School of Veterinary Medicine, another for Engineering, one for Music, one for Arts and Craftsmanship (i.e. technology), and one for Agriculture (Omar Al-Iskandarī and Salīm Ḥassan, 1917A).

In this new program of technical and vocational education, Mohammad Ali had to face two main interrelated problems : the hiring of teaching and technical staff, and the language of instruction. The teaching staff in medicine and surgery were a mix of Italian, French, Spanish, and German physicians and surgeons. Though most of these instructors were apparently quite good, a few turned out to be pretenders or opportunists with spurious credentials (Aida Nosseir, 1994A, 244). The wise leader also initiated a well-planned program for sending Egyptian students to Europe, with an eye toward gradually replacing imported experts with Egyptians.

Three small groups, totaling 28 students, were sent between 1813 and 1825 to Italy, France, and England to learn engineering, military subjects, ship building, navigation, hydraulics, and mechanics. While in Europe, one of these students became well versed in printing technology, and in 1821 became the manager of the highly influential Bulaq Press — the first real Egyptian Press. Later, nine large groups, totaling 291 students, were sent to Europe between 1826 and 1847 — mainly to France. These students were taught French before being sent abroad.

Members of the first large group (44 students) studied military management and engineering, artillery, law, political science, navigation, agriculture, irrigation, mining, medicine, surgery, and natural history. The religious leader of the group was Sheikh Rifat al-Ṭahṭawī, who eventually proved to be the brightest of the lot. Most members of the second large group studied mathematics and engineering, while those of the third were oriented toward acquiring various technological skills and the skills required to run small industries. The fourth group (12 students), sent in 1832, was known as the " major medical mission " since its students focused on medical training. In this way, each group specialized in a particular area of scientific and/or technical knowledge (Ibrahim Badran *et al.*, 1995A). Many of the graduates of these student missions became well-known figures in the great Egyptian intellectual resurgence or renaissance.

As for the language of instruction in Egypt, Dr Clot Bey appointed interpreters to translate lessons from French to Arabic, and encouraged students in the advanced classes to help the beginners. Mohammad Ali tried to persuade the French teachers to learn Arabic, but they refused. As an alternative, a two-pronged plan was introduced : the interpreters were asked to study medicine along with the students to gain a better understanding of what they were translating, and the students were directed to study French in a school affiliated with the School of Medicine. Sheikh Ṭahṭawī became a teacher in this school after his return from France.

However, the most effective solution of all was the translation of books into Arabic. When a book was translated it underwent two phases of editing : first it was reviewed for the scientific accuracy of the translation. Then the Arabic language of the text was corrected by scholars from al-Azhar, who were most probably responsible for the curious rhymed titles — in vogue at that time — of many of the translated books.

" منتهى الأغراض ، فى علم شفاء الأمراض " (1834) .
" التحفة الفاخرة ، فى هيئة الأعضاء الظاهرة " (1834) .
" القول الصريح ، فى علم التشريح " (1835) .

Some examples include : the first translations were from Italian, but most of these were never published. Later, French teachers and experts gained the confidence of Mohammad Ali, and French sources came to dominate. In the 1820's twenty-one books were translated. This jumped to 102 in the 1830's ; 75 % of these books were translated from French. English-language sources were very limited during Mohammad Ali's rule in the first half of the nineteenth century, but later they crept in gradually. By the 1880's and 1890's they were equal in number to the French sources (Aida Nosseir, 1994A, tables 47-55). A few books were translated to and from Turkish, as well. Most of these apparently addressed administrative and military matters. Turkish was still the official language in government offices until Ismaīl Pasha's decree in 1869 ordering that Arabic would be the official language of Egypt. By the end of the nineteenth century, books translated into Arabic totalled 677 (84.2 % of the total) while those translated into Turkish totaled 113 (14 %).

During Mohammad Ali's rule the books selected for translation were closely related to the courses taught in the schools and the immediate technological needs of developing industries. For example, Mohammad Ali encouraged the development of silkworm cultivation and an Egyptian silk-textile industry. Thus we find among the very early publications of the Bulaq Press a book on silk dying translated by Rafaēl Zakhor in 1823. Headmasters of the various schools were asked to make annual selections for translation of new books published in their fields. When books were not available in Egypt, al-Ṭahṭawī ordered them from Europe.

In the first half of the nineteenth century the subjects of translated books were predominantly in the applied and basic sciences, accounting for 56.3 % (147) of the 261 books translated. In the second half of the century, this percentage of the total fell by about half (27.2 %), though the raw number remained almost the same (148 books) because the total number of translated books rose, and interest in social sciences and humanities increased (see table 1). On average, 500-1000 copies were printed of each translated book.

Table 1 : Number and ratio of books in applied and basic sciences translated in the first and second halves of the nineteenth century in Egypt[2]

	1st half of the century		2nd half of the century	
Subject	Number	%	Number	%
Applied Sciences	91	34.9	68	12.5
Basic Sciences	56	21.4	80	14.7
Total	147	56.3	148	27.2
Total Translated	261		543	

At first books were translated by the interpreters from the medical and technical schools. Most of these were Syrians, notably Rafaēl Zakhor Rahibah, Yoḥanna 'Anḥourī and Yussuf Pharaon. Rafaēl Zakhor made an Italian-Arabic Dictionary, the first book to be published by the Bulaq press in 1822. His successor 'Anḥourī translated many books, but as he was not proficient in French, books were sometimes first translated for him from French to Italian, from which he made his Arabic translation. Pharaon's translations were related to veterinary medicine (Gmal al-Shayal, 1951A).

Graduates of the student missions to Europe soon replaced this first generation of translators, as Mohammad Ali had intended. However, Mohammad Ali did not reward translators with the weight of their translated works in gold, as the Caliph al-M'amūn is said to have done some eight centuries earlier. On the contrary, he was rather hard on them. Sometimes they were asked to translate the books they studied abroad even before returning home, as Sheikh al-Ṭahṭawī did. Some of them were asked to begin their translation while they were still confined in quarantine ! Occasionally they were kept behind the closed doors of the citadel until they finished the translation assigned to them. The presentation of a translated book was sometimes a prerequisite for employment (Jack Tajir, 1945A). To give a sense of the number of translators at work in Egypt in this period, Aida Nosseir (1994A, 301-306) gives a list of 61 translators in the first half of the nineteenth century and 114 in the second half.

The various state entities dealing with translation were eventually combined in the School of Translation, which was established in 1835 under the academic and administrative direction of Sheikh al-Ṭahṭawī. (The school later became known as the School of Languages, *Madrasat al-Alsūn*). This was an institution of teaching and training as well as the actual production of trans-

2. Adapted from table 61, p. 286 (Aida Nosseir, 1994A).

lated books. About 70 translators and writers graduated from the School of Translation between 1836 and 1847. In 1841 Sheikh al-Ṭahṭawī established an office (or *qalam*) for translation, which was later subdivided into offices for mathematics, natural and medical sciences, social sciences and humanities, and Turkish language.

However, in 1850 Abbas Pasha closed the School of Languages after 15 successful years of rich production. When Said Pasha came to power, al-Ṭahṭawī returned from the Sudan and established (in 1856) a new office for translation affiliated to a military school at the Citadel. Later, in the early years of the rule of Ismaīl Pasha, al-Ṭahṭawī established yet another new translation office (Aida Nosseir, 1994A). Thus this great translator, instructor and pioneer of enlightenment persistently struggled to bring new works to Egyptian scholars through translation.

By this time the Bulaq Press (established in 1820) was not the only one at work. Another was established in Saray Rās al-Tīn in Alexandria, and smaller presses were affiliated with some of the higher technical and military schools. The number of presses gradually increased, reaching 95 in the 1890s. The number of published works also steadily increased, totaling 867 in the first half of the nineteenth century and 9,538 in the second half. However, a large proportion of the books published in the second half of the nineteenth century (56.4 %) was for use in the public schools.

The ratio of works in basic and applied sciences to the total published in the first half of the nineteenth century was about 27 % (out of 867), but this fell to only about 9.5 % (out of 9,538) in the second half — even though the absolute number increased from 236 to 920 (see table 2).

Table 2 : Number of books in basic and applied sciences and their ratio to total works published in Egypt in the first and second halves of the nineteenth century[3]

	1st half of the century			2nd half of the century		
Subject	Number	%	Order	Number	%	Order
Applied Sciences	147	16.9	2nd	431	4.5	8th
Basic Sciences	89	10.2	6th	489	5.0	7th
Total	236	27.1		920	9.5	
Total Published	867			9,538		

It is significant that in the first half of the century books published on applied sciences came in second among the 10 fields of knowledge in the Dewey's decimal system, with about 17 % of all published works (147 books), while basic sciences ranked the sixth, with about 10 % (89 books). In the sec-

3. Adapted from tables 31 and 45 in Aida Nosseir (1994A).

ond half of the century basic and applied sciences became almost equal : the applied sciences fell from the second to the eighth rank, with only 4.5 % (431 books), while the basic sciences remained at almost same rank (seventh), with 5 % of the total (489 books).

Translated works formed 30 % of the total published in the first half of the nineteenth century, but dropped to about 6 % in the second half (though they doubled in total number). Naturally, the production of original Arabic books in science and technology was dependent on Egyptian scholars' access to information and ideas in these fields, and thus on the existence of translation and translated works.

Most of the scientific books translated into Arabic or written in Arabic in this period were of a high standard. One of their merits is their pioneering endeavour to coin Arabic scientific terms corresponding to the numerous foreign ones encountered in the literature.

Dr Mohammad Haytham al-Khayat states that later these terms were of great help in making the English-Arabic-French " Unified Medical Dictionary " adopted by WHO and ALECSO (M.H. Khayat, 1984A). Also, the range of subjects covered in the nineteenth century publications was extremely broad, as is evident in the comprehensive catalogues prepared by Y.I. Sarkīs (1928A) and Dr Aida Nosseir (1990A, 164-192).

A good example of the quality of publication coming out of Egypt at this time may be found in the field of biology, in a book titled *Tanwīr al-Adhhan* (*The Enlightenment of Minds*) by Dr Bishāra Zalzal. This book of 368 pages of small type was published in 1879, printed at Alexandria, dedicated in prose and poetry to Sultan Abdulhamid, and registered at the " Dignified Ministry of Education " in the " Exalted Asitāna ". The author stated that he had written the book for school pupils. The book is an excellent, well-illustrated, well-written, and meticulously documented work covering natural history, zoology and animal classification, anthropology, human biology, religion, sociology, and civilization. In the introduction, the author mentions al-Jāhiẓ, al-Damīrī, al-Kutbi, and al-Karmany together with Aristotle, Linnaeus, Buffon, Cuvier, Saint-Hilaire, Goethe and Lamarck, and discusses problems of good translation and writing science in Arabic. The sections on biology represent a first-rate exposition of the knowledge available at that time. They include discussion of comparative anatomy and physiology, cytology, histology, and embryology. The part devoted to animal taxonomy starts with chapters on Lamarckism, Darwinism, evidence of evolution, natural selection, and zoogeography. The discussion of Darwinism (p. 105-141) is particularly notable. It is remarkable to find such a comprehensive and up-to-date overview of this topic in a school textbook only twenty years after the first publication of Darwin's controversial " Origin of Species " — and 46 years before the famous " monkey trial " opened on July 20, 1925, where John Thomas Scopes, a young High School teacher in the American state of Tennessee was indicted for teaching evolution (Mohammad, 1994).

In addition to books, magazines and other periodicals were instrumental in the dissemination of modern scientific ideas among a wider sector of Egyptian society. Among the earliest and most important publications was *Al-Waqa'i' al-Miṣriya* (*The Egyptian Official Gazette*), which first appeared in 1828 — the oldest Arab newspaper. When it first appeared the *Gazette* was written in both Arabic and Turkish and contained local and foreign news, as well as official decrees and announcements. In 1830 Sheikh Ṭahṭawī became its editor. Over the course of the nineteenth century the number of newspapers gradually increased, and by the end of the century there were 26 newspapers and 37 magazines published in Arabic. Some of these were particularly concerned with scientific, agricultural, medical, and general cultural subjects (Aida Nosseir, 1994A).

Two monthly magazines are of particular relevance to our present discussion : *al-Moqtaṭaf* (meaning " anthology, harvest or selection "), and *al-Hilāl* (*The Crescent*). *Al- Moqtataf* was founded in Beirut in 1876 by Dr Yaqūb Ṣarrūf and Dr Fāris Nimr, but was later transferred to Cairo in 1885, where it continued to appear until 1950. In a symposium organized by IRCICA in Istanbul in 1994, I reviewed in some detail articles, news, and discussions concerning organic evolution that had been published in its 116 volumes discussing the reception of Darwinism in some Muslim countries (A. Helmy Mohammad, 1994). In general, there was a marked inclination in *al-Moqtataf* toward scientific subjects, since Ṣarrūf and Nimr had been teaching natural history in the American University in Beirut before coming to Egypt (Ahmad Hussein al-Sawi, 1992A).

Al-Hilāl, founded by Georgi Zeidan in Cairo 1892, was named in honor of the Ottoman crescent, and also to evoke the image of the monthly appearance of the new moon in the night sky — a symbol of hopeful development to a full moon (A.H. al-Ṣawi, 1992). Though Zeidan attempted to study medicine in Beirut (1880-1881) and in Cairo (1883), he eventually became a journalist, professional translator and novelist. *Al-Hilal* tended toward literary subjects, but interest in scientific topics was by no means weak. Between 1892 and 1914 articles appeared on the following topics : various aspects of organic evolution and Darwin's ideas, heredity and eugenics, anthropology, extinct reptiles, an obituary of Louis Pasteur, X-rays and radio-active elements, various animals, birds and insects, agriculture, cancer research, medicine, public health and nutrition, many articles on astronomical subjects, magnetism, oceanography, life and death, and epidemiology of microbial diseases.

Not surprisingly, the far-reaching and well-organized plan of science and technology transfer that reached its peak in Mohammad Ali's time was intimately related to a parallel wave of progress in economical, military, agricultural, technological, and industrial fields. It will suffice here to mention only a few important features of this boom. New excellent strains of cotton were introduced in 1820, ushering in a new era defined by a cotton-based agricultural economy and related industries. Silkworms were also raised and mulberry trees cultivated, creating a silk-textile industry that yielded 10,000 kilograms of silk in 1833. Programs of sugar-cane cultivation and sugar purification also produced very good results. Canals from the Nile were cut for irrigation and water transport. In 1834 studies were initiated for building a strong barrier before the bifurcation of the Nile north to Cairo, and actual construction of this barrage began in 1843.

All this was ultimately in support of — and supported by — a strong army and powerful naval fleet. These demanded active mining, metallurgy, gunpowder production, other military engineering crafts, and the establishment of high standard arsenals. The arsenal at Bulaq constructed the Red-Sea fleet. Meanwhile, Alexandria built the Mediterranean fleet and restored it after its destruction at Navarino Bay in 1827. Some of these industries deteriorated after Mohammad Ali’s time, but others were sustained and even came to flourish, especially during Ismaīl Pasha’s rule (Hassan Ismail *et al.*, 1993 ; Omar al-Iskandari and Salīm Hassan, 1917).

In conclusion, I want to bring out the real significance and value of the transfer of modern science and technology to Egypt by way of some “ flashbacks ” to the great era of the transfer of scientific knowledge from Antiquity to the rising early Islamic civilisation. In Mohammad Ali’s time the major routes of transfer and exchange were translation, the importation of human expertise from Europe, and the sending of student missions to Europe. In the days of al-Ma’mūn, translation was the main, indeed almost the only, route. Another difference is that the earlier era of translation was from an almost closed entity (Classical Antiquity), a heritage that could be revived and developed without any real competition with its authors. In the nineteenth century translation was — as it still is — an activity that took place in dialogue with contemporary and actively developing sources.

A third difference is that the early Muslim recipients of the Classical heritage were on the rise as a civilization, whereas recipients of the influx of Western scientific knowledge in the nineteenth century were in a subordinate position. They were trying to wake up to a new world and intellectual order, and to avoid getting entangled in successive waves of dominating foreign intruders. Professor Ihsanoğlu brings to our attention another basic point related to the types of translated works seen in each period. His thesis is that the early Muslims translated important works of basic science, while their later successors were after sources addressing urgently-felt demands for applied science (i.e. technology). His most significant examples of this trend have been drawn from the field of astronomy (Ihsanoğlu, 1988A).

Finally, I would like to offer a rather vague idea which I would like to see worked out : Early Muslims represented a new-born community with a new universal message. They were revolting against old traditions, but tolerant of and even welcoming transfer from earlier civilisations and contact with other people, irrespective of any ethnic barrier. The new faith commanded its adherents to seek knowledge (*'ilm*) from any source and by all means (A. Helmy Mohammad, 1997A). Late Muslims, on the other hand, have a historical burden of constraints and problems and have somehow lost a clear sight of the basic principles of their religion and their initial zeal to follow them. History provides roots and a deep anchorage in time, but — in a way — can also be cumbersome.

A few weeks ago at Amman's Academy of Arabic Language we were discussing a certain aspect of this problem : the importance of the movement toward Arabic as the language of instruction in Egyptian and other Arab universities (A. Helmy Mohammad, 1997bA). We recalled that early in the nineteenth century the Japanese were dazzled by the Egyptian renaissance and sent envoys to study it as a model. Later, the Meiji revolution in Japan followed along similar lines to achieve the Japanese scientific and technological triumph we are all familiar with. Some Arabic scholars, including Dr Mohammad Haytham al-Khayat (1984A) argue that Egypt was once positioned to make the same leap, but failed to achieve the same results because in 1887, after almost 60 years of instruction in Arabic (since the early efforts of Mohammad Ali), the British decreed that the language of instruction in Egyptian schools of higher education should be English.

Nationalization of the language of science teaching in universities and other institutions of higher education is undoubtedly an important element in the real progress of a developing nation (A. Helmy Mohammad, 1997bA). However, many other factors, such as those I have discussed here, are also intricately involved. In particular, we should attend to the socio-economic and socio-political dimensions of this problem.

References

E. Ihsanoğlu (ed.), " Transfer of Modern Science and Technology to the Muslim World ", *Proceedings of the International Symposium on Modern Science and the Muslim World (Istanbul, September 2-4, 1987)*, Istanbul, IRCICA, 1992.

A.H. Helmy Mohammad, " Casual Notes on the Reception of Darwinism in Same Islamic Countries ", *Proceedings of the symposium " Science and Technology in the Muslim World " (Istanbul, June 3-5, 1994)*, Istanbul, IRCICA (in press).

S.H. Nasr, *Islamic Science : An Illustrated Study*, Kent, World of Islam Festival Publishing Company Ltd., 1976.

F. Robinson, *The Cambridge Illustrated History of the Islamic World*. Cambridge, Cambridge University Press, 1996.

إبراهيم جميل بدران وآخرون ،١٩٩٥. " تأريخ الحركة العلمية فى مصر الحديثة : العلوم الطبية. أكاديمية البحث العلمى والتكنولوجيا ، القاهرة (٦٢٤ ص).

أحمد حسين الصاوى،١٩٩٢. " الهلال : مائة عام من التنوير والتحديث " . دار الهلال ، القاهرة.

أحمد حسين الصاوى وآخرون،١٩٨٠ . " كشاف الهلال ، الجزء الأول : سبتمبر ١٨٩٢- يوليو ١٩١٤ " . دار الهلال ، القاهرة .

أكمل الدين إحسان أوغلى ،١٩٨٨ . " دخول العلوم الحديثة إلى العالم الإسلامى " . محاضرات الموسم الثقافى : ٢٤٢-٢٧٩ . مؤسسة الثقافة والفنون ، أبو ظبى .

بشارة زلزل،١٨٧٩ . " كتاب تنوير الأذهان : فى علم حياة الحيوان والإنسان وتفاوت الأمم فى المدنية والعمران " . مسجل فى نظارة المعارف الجليلة فى الآستانة العلية ، مطبعة مجلة الجامعة بالاسكندرية .(٣٦٨ ص) .

جاك تاجر،١٨٤٥. " حركة الترجمة بمصر خلال القرن التاسع عشر" . دار المعارف ، القاهرة (١٥٨ ص) - (عن : عايدة نصير ،١٩٩٤) .

جمال الدين الشيال،١٩٥١. " تأريخ الترجمة والحركة الثقافية فى عصر محمد على" . دار الفكر العربى ، القاهرة. (٢٢٨ ص) - (عن : عايدة نصير ،١٩٩٤) .

حسن إسماعيل وآخرون،١٩٩٣. " تأريخ العلوم والتكنولوجيا الهندسية فى مصر ، فى القرنين التاسع عشر والعشرين" . أكاديمية البحث العلمى والتكنولوجيا ،القاهرة. (جزءان ،١٠٤٦ ص) .

عايدة إبراهيم نصير،١٩٩٠ ." الكتب العربية التى نشرت فى مصر فى القرن التاسع عشر" . قسم النشر بالجامعة الأمريكية بالقاهرة . (٤٠٣ ص) .

عايدة إبراهيم نصير،١٩٩٤ ." حركة نشر الكتب فى مصر فى القرن التاسع عشر" . الهيئة المصرية العامة للكتاب، القاهرة . (٦٥٤ ص) .

عبدالحافظ حلمي محمد،١٩٩٧– a . " الإسلام واللغة العربية والعلم " . المؤتمر السنوى لمجمع اللغة العربية بالقاهرة فى دورته الثالثة والستين" . (تحت الطبع)

عبدالحافظ حلمي محمد،١٩٩٧ – b . " اللغة العربية وتعريب التعليم الجامعى فى مصر فى الحاضر ونظرة إلى المستقبل : قضايا وحلول" . الموسم الثقافى الخامس عشر لمجمع اللغة العربية الأردنى" . (تحت الطبع) .

عمر الإسكندرى،وسليم حسن،١٩١٧. " تاريخ مصر من الفتح العثمانى " (مراجعة : أ.ج.سَفِدْج) . مطبعة المعارف ، القاهرة . (٣٠٤ ص) .

محمد قنديل البقلى،١٩٥٨. " المختار من تاريخ الجبرتى " . مطابع الشعب ، القاهرة . (تسعة أجزاء ، ١٠٦٨ ص) .

محمد هيثم الخياط، ١٩٨٤. " تعريب العلوم الطبية " . الموسم الثقافى الثانى لمجمع اللغة العربية الأردنى "(٢١ ص) .

المقتطف(دورية شهرية) ، ١٨٧٦–١٩٥٠ . بيروت – القاهرة .

الهلال (دورية شهرية) ، ١٨٩٢– الوقت الحاضر . دار الهلال ، القاهرة .

يوسف إليان سركيس،١٩٢٨. " معجم المطبوعات العربية والمعربة " . مطبعة سركيس ، القاهرة . (مجلدان ، ٢٠٢٤ عموداً) .

An Ottoman Professor of Botany : Salih Efendi (1817-1895) and his Contributions to Botanical Education in Turkey

Feza Günergun

Although botany was not included in the curriculum of the *medrese* — the classical Ottoman institution of learning and education — the Ottomans were acquainted with botanical knowledge, especially concerning the properties of medicinal plants. Ottoman scholars came to this knowledge through translations based on Ibn el-Baithar's *El-Müfredat* (*Traité des Simples*), Dioscorides' *Materia Medica,* or Avicenna's *Kitab el-Nebat* (*Book of Plants*).

The earliest Turkish printed text on botany is, as far as we know, a text published in the fourth final volume (1834) of *Mecmua-i Ulum-i Riyaziye* (*Compendium of Mathematical Sciences*) by Ishak Efendi, chief instructor of the *Mühendishane-i Berri-i Hümayun* (Imperial School of Engineering). In this text, plants are classified as trees, shrubs, and herbs. Attention is given to the root and its function, and to the concentric annual rings in woody stems. Floral structure is also described. Plant reproduction by seeds, water circulation, nutrition, respiration, and photosynthesis in green plants are explained. Edaphic conditions for a favourable utilization of soil water and nutrients, and the damage caused by overwatering are discussed. In short, we find here a general overview of the fundemental subjects of botany and their practical application. As botany was not part of the curriculum of the Imperial School of Engineering, the presence of botanical information in a text-book prepared for engineering students may be explained by the author's aspiration to compile an encyclopedic work dealing with various sciences.

The teaching of botany started in Turkey in the first half of the nineteenth century as a component of medical education. In the early years of the *Tiphane-i Amire* (State School of Medicine, est. 1827), the Turkish and Arabic names of plants were taught to first-year students. By 1834, a course titled *Ilm-i Nebat* (Botany) had been added to the program of the fourth year. In the *Mekteb-i Tibbiye-i Shahane* (Imperial School of Medicine), established in

1839, botany was initially taught in the first year by the Austrian physician Dr. C.A. Bernard, who was also the director of the school. As the language of instruction at the *Mekteb-i Tibbiye* was French, botany courses were also given in French, and Dr. Bernard's book *Elémens de Botanique* (1842) was used as a textbook. This book (*ca.* 360 pages) which was written in French and follows A. Richard's system of classification was the first textbook of medical botany published in Turkey. Dr Bernard wrote also a book on the thermal springs of Bursa, *Les Bains de Brousse* (1842) in which he added notes on the vegetation of Uludag and of the route from Bursa to Mudanya. He collected plant specimens in these locations and listed their names. Another contribution of Dr Bernard to the botanical education in Turkey was the establishment of a botanical garden in the grounds of the Imperial School of Medicine, from which he initiated practical courses on botany by using the living material grown in this garden.

Dr Bernard had a meticulous and diligent student to assist him in teaching botany. This was Salih Efendi (1816-1895). Indeed, Salih Efendi started teaching *Ilm-i Nebatat* (Botany) even before graduating, and took part in the establishment of a botanical garden at the Imperial School of Medicine. In 1843 he was among the first graduates of that school, and received the title of *Docteur en Médecine et en Chirurgie*. As Dr Bernard noted in the report he presented to the Sultan on the activities of the academic year 1842-43, Salih Efendi, due to his talent and careful study, would soon become one of the most productive and well-known members of the teaching staff. And indeed, after the death of C.A. Bernard in 1844, Salih Efendi was appointed to teach botany at the Imperial School of Medicine.

In addition to his post as professor at the Imperial School of Medicine, Salih Efendi was appointed " Hekimbasi " (Chief Physician) to the Sultan. He was also appointed twice to the directorship of the Imperial School of Medicine, in the years 1849 and 1865. He served as a councillor to the Ministries of Education and Trade, and served as member and chairman of various state offices and assemblies. He chaired the International Health Congress held in Istanbul in 1865. He received the " Bâlâ " degree and was presented with medals of honour by foreign governments including France, Prussia, Portugal, and Spain.

Despite his several administrative posts, Salih Efendi never gave up teaching botany. He translated a book on natural sciences in 1865, and prepared a second edition in 1872. After retiring from the Imperial School of Medicine, he continued to lecture on botany for many years at the *Mekteb-i Tibbiye-i Mülkiye* (Civilian School of Medicine). He gave natural history courses at *Darülmuallimin* and *Darülfünun* (Ottoman institutions founded for training, respectively, teachers and the general public). He designed the garden at his house in the Istanbul suburb of Anadoluhisari as a botanical garden where he cultivated various kinds of plants and fruit trees. He died in 1895 in this same house, which is still one of the most beautiful waterfront residences on the Bosphorus.

The aim of this paper is twofold. The first aim is to introduce Salih Efendi, an Ottoman physician and buraucrat known for his keen interest in plants and his contributions to the teaching of modern botany in Ottoman Turkey. The second aim is to review briefly the botanical section of Salih Efendi's book *Ilm-i Hayvanat ve Nebatat* (*Zoology and Botany*, first printed in 1865) to attempt to discover the source from which it was translated. Finally we will bring to light this book's second modified edition, published in 1872.

ILM-I HAYVANAT VE NEBATAT (*ZOOLOGY AND BOTANY*, 1865)

As far as we know, the botanical section of Salih Efendi's *Ilm-i Hayvanat ve Nebatat* was the first illustrated text on systematic botany published in Turkish. It was certainly the first book in Turkish used in the teaching of botany. Although there was already a 64-page botanical text in the *Mecmua-i Ulum-i Riyaziye* (*Compendium of Mathematical Sciences*, vol. 4, 1834), compiled from European sources by the chief instructor of the Imperial School of Engineering Ishak Efendi, this text contained no illustrations and dealt with practical and physiological botany rather than systematic botany. It was probably not used in teaching botany. On the other hand, C.A. Bernard's above mentioned textbook, *Eléments de Botanique* (1842), dealt with medical botany, rather than systematic botany.

In the foreword to *Ilm-i Hayvanat ve Nebatat*, Salih Efendi explains that he translated the book into Turkish for general use, but particularly to be taught at the *Rüsdiye* (secondary) schools. In fact, the original book he translated had been written for the same purpose — that is " to be used by young people ". Salih Efendi was a member of the Council of Public Education, which was founded in 1841 to improve the Empire's educational institutions, including secondary schools. It may have been this duty that prompted him to translate a natural history book to be used in these schools. Furthermore, if we consider that Salih Efendi had been teaching botany at the newly-established Civilian School of Medicine (est. 1867), and that no other botany books were available in Turkish before 1872, it can be argued that his book was to be used as an elementary textbook in the botany courses given at the medical school. On the other hand, we also know that in 1863 Salih Efendi was invited to give lectures on *Ilm-i Mevalid* (Natural History) at the newly established *Darülfünûn*, the Ottoman institution of higher education founded in 1863 with the aim of creating an university. He taught natural history and related subjects there from 1863 to 1865, and in 1865 his book was published. Thus it is very probable that Salih Efendi used *Ilm-i Hayvanat ve Nebatat* in the natural history courses he gave at the Darülfünun. Indeed, in the foreword to the book, he states that he lectured natural history to the general public in the *Darülfünun* at the request of the Sultan, and that he translated the book for common usage. The same argument is encountered in the petition that Salih Efendi addressed to the Sultan in 1866 when presenting the book : " I translated this work after I was appointed to the *Darülfünun* as lecturer, and considered that the propagation of natural sciences in the Ottoman realm would be beneficial ".

Ilm-i Hayvanat ve Nebatat (1865) consists of 90 pages of text plus 25 plates with illustrations. As Salih Efendi explains in the foreword, the book contains a lot of illustrations and the text is rather short, for it was intended for general use, and especially as a textbook for secondary schools. Fifty four pages (p. 4-57) are devoted to the zoology, and the last 32 pages (p. 58-89) deal with botany. Appended to the text are 18 plates (plates 1-18, including 184 figures) on animals and 7 plates (plates 19-25, with 106 figures) on plants. The total number of figures in the plates is 290. Under each figure, the Latin name of the plant is given in both Latin and Arabic characters together with its Turkish name and Turkish family name. When figures of plant sections are added, these are also explained.

The book does not contain a cover page bearing its title, the name of the author, the printing house, and the publication date. The title of the book and the name of its author are mentioned in the foreword, the publication date (24 November 1865) is given at the end of the text. The heading " Ilm-i Hayvanat ve Nebatat " on the first page has been accepted as the title of the book.

After a brief introduction where zoology, botany and mineralogy are discussed comes a chapter on general botany where cells, tissues, the vegetative organs (root, stem and leaves) and sexual organs are discussed. The French names (in Latin characters) of these organs and tissues are given together with their Turkish equivalents. The number of French terms in the whole book is around 70. Next (p. 69-89), artificial and natural classifications of plants are discussed and the systems of Linnaeus and Jussieu are introduced.

Linnaeus' system is introduced as an artificial classification based on the number of stamen in plant's flowers. Plants are grouped into 24 classes. The Latin and Turkish names of these classes are indicated and their characteristics explained. The Turkish and Latin names of various plants belonging to these classes are given. Approximately 90 of these plants are defined by their specific names, and 40 of them by their generic names.

In Jussieu's system of natural classification, plants are first divided into three major classes according to their cotyledon numbers and then into minor classes. There are a total of 15 classes. In addition to the general characteristics of each class, Turkish and Latin names of plants exemplifying these classes are given. The names of the classes (*sinif*) and divisions (*firka*) are written in Turkish, with the French equivalents given in brackets. This section contains the names of 25 higher categories, 44 species names and 1 genus name.

After this brief description of the *Ilm-i Hayvanat ve Nebatat*, let us consider the origin of the book. Salih Efendi states in the foreword that he prepared the book after the " French edition of a treatise on natural sciences written by a Bavarian physician and instructor called Arns ". According to this statement, the original book should have been on zoology, botany, and mineralogy, and should have been translated into French from German ; and the French edition shoud have a date prior to 1865. Research we made at Heidelberg University Library in June 1995, revealed that the book translated by Salih Efendi was Dr Carl Arendts' (1815-1881) *Eléments d'Histoire et de Technologie à l'Usage de la Jeunesse.* The latter was translated into French by a Dr Royer from the *Arendts' Naturhistorischer Schulatlas.* The French edition, published in Brussels and Leipzig in 1859, consisted of 5 sections : *Zoologie, Botanique, Minéralogie, Technologie, Géognosie.* 33 plates bearing 388 woodcut illustrations were appended to the 44-page text.

When Arendts' French edition is compared to *Ilm-i Hayvanat ve Nebatat,* it is clear that Salih Efendi translated only the zoology and botany sections and added the corresponding 25 plates to the end his translation. Plate and figure numbers in the text were kept as they were. Naturally, the explanations under the figures were translated into Turkish.

USUL-I MENAKIB-I TABIIYAT (*ELEMENTS OF NATURAL HISTORY*, 1872)

Ilm-i Hayvanat ve Nebatat (1865) was republished seven years later in 1872, in Istanbul. The 1872 edition starts off with a chapter titled " Usul-i Menakib-i Tabiiyat ", or " natural sciences ". Consequently this title was entered into library and book catalogues as the title of the book, without realizing that this was a new edition of *Ilm-i Hayvanat ve Nebatat* (1865), and was regarded as a distinct work by Salih Efendi. Thus we introduce here this second book of Salih Efendi's, dated 1872, to the botany literature as the second edition of *Ilm-i Hayvanat ve Nebatat.*

When the two editions are compared, the first point that strikes the reader is that all Latin and French botanical terms written in Latin characters in the first edition are eliminated in the second. For example, while the Latin equivalent for " zagferan " (a bulbous plant) is found in the 1865 edition as " crocus autumnalis ", this Latin name is removed in the 1872 edition. Secondly, there is a preference for Turkish syntax instead of Arabic or Persian. For example, Salih Efendi writes *hayvanatin cismi* instead of *cism-i hayvanat,* both expressions meaning animal body.

The removal of the French and Latin equivalents of Ottoman botanic terms from the text by Salih Efendi should be evaluated within the attempts undertaken by Turkish physicians of this time to establish Turkish as the language of medical education. The number of graduate physicians from the Imperial School of Medicine (Military School of Medicine), where instruction had been in French since 1839, was not sufficient to meet the needs of the Empire and thus the teaching of medicine in Turkish was planned. With the efforts of Turkish professors, a separate school called the Civilian School of Medicine, where the instruction would be in Turkish, was established in 1867 within the Imperial School of Medicine. The founders firmly defended the argument that medical education could be and should be given in Turkish. Strong objections were raised by francophone professors in 1870 when it was decided that instruction in Turkish should also start at the Imperial School of Medicine. The second edition of Salih Efendi's book came out just at the time of these serious discussions between Turcophone and Francophone physicians. Instruction in Turkish was finally adopted in the military medical school thanks to the conviction of Turkish teaching staff. The elimination of French and Latin terms by Salih Efendi reflects the support he brought to the project of teaching medicine in Turkish.

Conclusion

Salih Efendi's career reflects perfectly the cultural and scientific transformations which occurred in the Ottoman world in the 19th century. Born in the second decade of this century, he was probably educated in the *medrese*, the classical Ottoman institution of learning, where he excelled in Arabic. He studied in the Imperial School of Medicine, founded in 1839, where the language of instruction was French. Due to his interest in and talent for botany, he started to teach in this field while he was a student at the medical school. He was appointed as professor of botany as soon as he graduated. He also lectured on botany in the Civilian School of Medicine for many years. In his scholarly and administrative careers, he played an important role in the introduction of modern sciences in the Ottoman World. His duties in these educational institutions, as well as the lack of a botany book in Turkish, led him to translate on the principles of botany. The translation he made from C. Arendts' book on natural history was published in Istanbul in 1865. It contains the first Turkish text on systematic botany. The second edition, published in 1872, represents the change that had occurred in the teaching language in medical education, where Turkish replaced French. In coining new scientific terms, Salih Efendi followed his contemporaries by using Arabic vocabulary. Thus Salih Efendi's career is a concrete example of an intellectual era in which Islamic and European scientific and educational traditions intermingled.

Acknowledgements

I would like here to thank Dr Edith Gülçin Ambros (Vienna), who sent photocopies from the French translation of Carl Arendts' *Eléments d'Histoire Naturelle et de Technologie* ; Doz. Dr Horst Remane (Leipzig) for providing photocopies of the zoology and botany sections of the above-mentioned French edition ; Dr Richard Lorch (Munich) for sending a biography of Arendts ; and Prof Dr Ekmeleddin Ihsanoglu for providing access to the data of the OMETAR project by IRCICA, thus enabling us to find Salih Efendi's petition to Sultan Abdülaziz in the Ottoman State Archives.

BIBLIOGRAPHY

1. " Ecole Impériale de Médecine de Galata-Sérai ", *Journal de Constantinople et des Intérêts Orientaux*, 1re Année, n° 51, Jeudi 21 Septembre 1843.
2. C.A. Bernard, " Rapport sur les Travaux de l'Ecole de Médecine de Galata-Sérai pendant l'Année Scolaire 1258/59, Présenté à sa Hautesse le 25 Schaban (20 Septembre) ", *Journal de Constantinople et des Intérêts Orientaux*, 1re Année, n° 52, Mardi 26 Septembre 1843. For a facsimile and the Turkish translation of this report see Semavi Eyice, " Dr Karl Ambros Bernard (Charles Ambroise Bernard) ve Mekteb-i Tibbiye-i Adliye-i Sahane'ye Dair Birkaç Not ", *Türk Tibbinin Batililasmasi* (Gülhane'nin 90. Kurulus Yildönümü Anisina 11-15 Mart 1988' de Ankara ve Istanbul'da Yapilan Sempozyuma Sunulan Bildiriler), ed. A.Terzioglu and E.Lucius (Istanbul, 1993), 97-124.
3. " Salih Efendi ", *Servet-i Fünun*, n° 214, 6 Nisan 1311/18 Nisan 1895, 82-83.
4. " Nécrologie ", *Revue Médico-Pharmaceutique*, 8e Annéee, n° 5, 31 Mai 1895, 79.
5. Mehmed Süreyya, " Salih Efendi ", *Sicill-i Osmani*, vol. III (Istanbul, 1311/1895), 219.
6. Besim Ömer, *Nevsal-i Afiyet*, vol. II (Istanbul, 1316/1898), 115-117.
7. Ibrahim Alaattin (Gövsa), *Meshur Adamlar, Hayatlari, Eserleri*, vol. IV (Istanbul, 1933-1936), 412.
8. Suheyl Ünver, " Eski Hekimbasilar Listesi (Hekim Hayrullah Efendi'ye göre) ", *Türk Tib Tarihi Arkivi*, *5*, n° 17 (1940), 7.
9. Saffet Eren, " Hekimbasi Salih Efendi Hakkinda, 1231-1312 (1816-1895) ", *Türk Tib Tarihi Arkivi*, *6*, n° 21-22 (1943), 9-23.
10. Ibrahim Alaattin Gövsa, *Türk Meshurlari Ansiklopedisi* (Istanbul, 1946), 342.
11. Süheyl Ünver, " Türkiye'de Nebatat Bagçeleri Tarihi Üzerine Küçük Bir Muhtira ", *Cerrahpasa Tip Fakültesi Dergisi*, *2*, n° 3 (1971), 449-553.
12. Mehmed Tahir (Bursali), *Osmanli Müellifleri*, ed. I. Özen, vol. III (Istanbul, 1975), 231.
13. Ekrem Kadri Unat, " Türk Tibbiye Mektepleri Muallimleri ", *Yeni Symposium*, *22*, n° 3-4 (1984), 23.
14. Turhan Baytop, *Türk Eczacilik Tarihi*, Istanbul, Istanbul Üniversitesi Eczacilik Fakültesi, 1985, 434.

15. Nil Sari, " Cemiyet-i Tibbiye-i Osmaniyye ve Tip Dilinin Türkçelesmesi Akimi ", *Osmanli Ilmi ve Mesleki Cemiyetleri,* ed. E. Ihsanoglu, Istanbul, Istanbul Üniversitesi Edebiyat Fakültesi, 1987, 121-142.
16. Ekmeleddin Ihsanoglu, " Darülfünûn Tarihçesine Giris : Ilk Iki Tesebbüs ", *Belleten, 54,* n° 210 (1990), 708, 709, 712.
17. Ekrem Kadri Unat ve Mustafa Samasti, *Mekteb-i Tibbiye-i Mülkiye (Sivil Tip Mektebi) 1867-1908*, Istanbul, Istanbul Üniversitesi Tip Fakültesi, 1990, 4, 12, 15, 17, 22.
18. Rengin Dramur, " Hekimbasi Salih Efendi (1816-1895) ", *Tip Tarihi Arastirmalari*, n° 4 (1990), 120-127.
19. Necdet Isli, " Tip tarihimizle ilgili birkaç mezar kitabesi IV ", *Tip Tarihi Arastirmalari*, n° 4 (1990), 114-118.
20. Riza Tahsin, Elhac, Binbasi, *Tip Fakültesi Tarihçesi (Mirat-i Mekteb-i Tibbiye)*, ed. Aykut Kazancigil, vol. I, Istanbul, 1991, 17, 28, 33, 36, 38, 80, 137-145.
21. Asuman Baytop, " 1839-1960 Yillari Arasinda Istanbul'da Basilmis Farmasötik Botanik Ders Kitaplari ", *Marmara Üniversitesi Eczacilik Dergisi, 8*, n° 1 (1992), 65-84.
22. Ayten Altintas, " Osmanli Imparatorlugunda Hekimbasiligin Lagvi Meselesi ", *Tip Tarihi Arastirmalari*, n° 5 (1993), 54-55.
23. Arslan Terzioglu, " Hekimbasi Salih Efendi ve Onun Joseph Hyrtl'e Yazdigi Fransizca Bir Mektup ", *Tarih ve Toplum*, n° 118 (1993), 30-36.
24. Nuran Yildirim, " Salih Efendi ", *Istanbul Ansiklopedisi*, vol. VI (Istanbul, 1994), 426-427.
25. Erol Özbilgen, " Bati Bilimini Türkiye'ye Aktaran Ilk Ders Kitaplari (3) ", *Müteferrika Bahar*, n° 5 (1995), 191-197.
26. Feza Günergun, Asuman Baytop, " Türkiye'de Modern Botanik Egitiminin Baslangici ve Dr C.A. Bernard'in Katkilari ", *Türk Tip Tarihi Yilligi (Acta Turcica Historiae Medicinae)*, vol. II, ed. A. Terzioglu, Istanbul, 1995, 135-152.

FIGURES

علم حیوانات ونباتات

محامد خلایق و برایا وتحیات موالید وذرایا اول فیض انوار
اقدسیله ممتنع التشکیل والتنظیم وعدیم التصویر والترسیم
اولان اجناس موجودات ومخلوقاتی برمفحوای (انا خلقنا کل
شیء بقدر) ابداع وانواع حیوانات ونباتاتك بقای تناسللرنده
ازدواج وتولیدی برمنطوق (انا خلقناکم من ذکر وانثی
ومامن دابة فی الارض ولاطائر یطیر بجناحیه الاامم امثالکم
وانبتنا فیها من کل زوج بهیج) ابداع بیوران ذات فرد متعال
تقدس عن الشبیه والمثال حضرتلرینه مخصوص واحرادر
وصلوات نامیه وتسلیمات زاکیه اول کل صدبرك ریاض
فیوضات تکوینات وزهرهٔ ثمارکلزار انوار تجلیات تعینات
زبدهٔ جملهٔ عالم سید ولد ادم هادیٔ سبل خاتم رسل
افندمز حضرتلرینه مستحق وسزادر وترضیهٔ وافیه غواص
بحار علم وعرفان وکاشف معادن یقین واذعان اولان آل
واصحاب اخیاره اهدا اولنورکه هربری لآلیٔ اصداف فنون
وحکمتی آفاق واقطاره نشر ونثار ایله طالبان معرفت
حقایقی بهره دار ایتدیلر * امابعد مکتب فنون طبیه نظارتیله
مباهی عبد حقیر صالح قوللری دارفنونده فن تاریخ طبیعی یی
باارادهٔ سنیهٔ حضرت پادشاهی عمومه تدریس ایتمکده

1. Salih Efendi, *Ilm-i Hayvanat ve Nebatat* (*Zoology and Botany*, 1865), first page

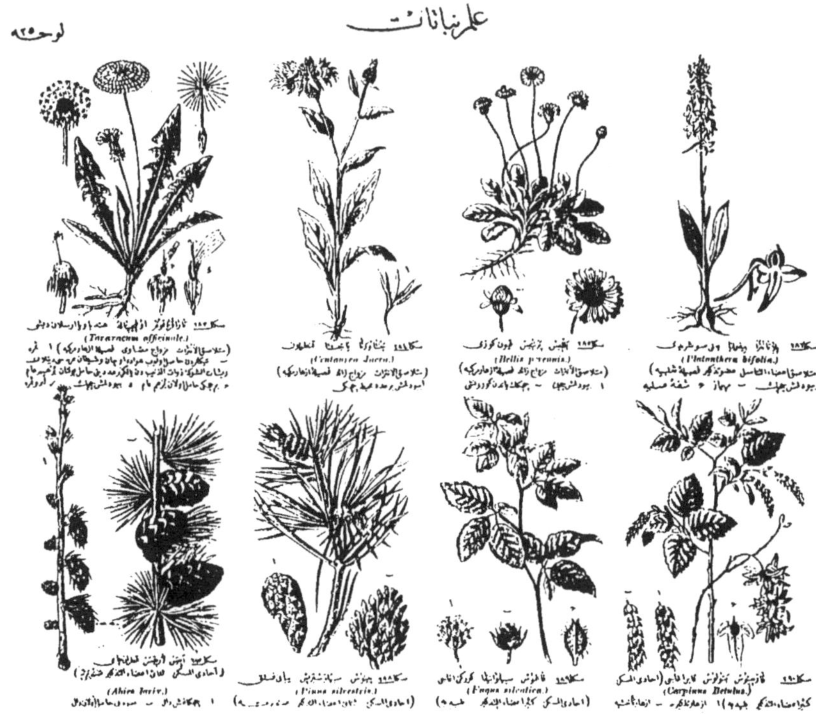

2. Salih Efendi, *Ilm-i Hayvanat ve Nebatat* (*Zoology and Botany*, 1865), plate 25

ÉLÉMENTS
D'HISTOIRE NATURELLE
ET DE
TECHNOLOGIE

A L'USAGE DE LA JEUNESSE

PAR LE

DR. CARL ARENDTS

PROFESSEUR AU CORPS ROYAL DES CADETS DE BAVIÈRE, MEMBRE DE PLUSIEURS SOCIÉTÉS SAVANTES.

Ouvrage enrichi de 33 tables et de 388 gravures sur bois

Avec le texte explicatif

traduit de l'allemand

par le

DR. P. ROYER

LEIPZIG
F. A. BROCKHAUS

BRUXELLES
AUGUSTE SCHNÉE
RUE ROYALE IMPASSE DU PARC 2

1859

3. C.Arendts' *Eléments d'Histoire Naturelle et de Technologie* translated by P. Royer, 1859, front page

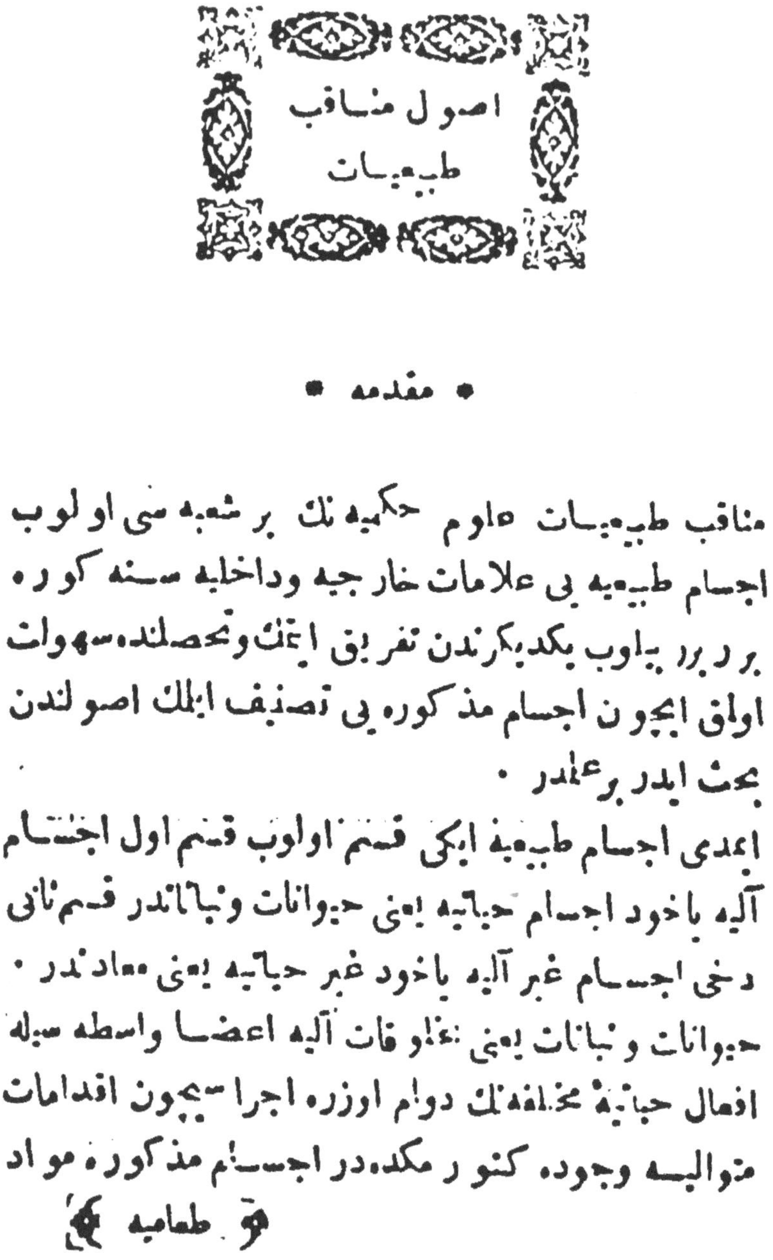

اصول مناقب
طبیعیات

* مقدمه *

مناقب طبیعیات علوم حکمیه نك بر شعبه سی اولوب
اجسام طبیعیه یی علامات خارجیه وداخلیه سنه كوره
بر بر بیلاوب یكدیكرندن تفریق ایتمك وتحصیلنده سهولت
اولق ایچون اجسام مذكوره یی تصنیف ایلمك اصولندن
بحث ایدر بر علمدر .
ایمدی اجسام طبیعیه ایکی قسم اولوب قسم اول اجسام
آلیه یاخود اجسام حیاتیه یعنی حیوانات ونباتاتدر قسم ثانی
دخی اجسام غیر آلیه یاخود غیر حیاتیه یعنی معادندر .
حیوانات و نباتات یعنی غذاوقات آلیه اعضا واسطه سیله
افعال حیاتیه محافظه نك دوام اوزره اجراسیچون اقدامات
متوالیه وجوده كتورمكده در اجسام مذكوره مواد
طعامیه

4. Salih Efendi, *Usul-i Menakib-i Tabiiyat* (*Natural History*, 1872), the 2nd edition of *Ilm-i Hayvanat ve Nebatat* (1865), first page

The Imperial Ottoman Izmir-to-Aydin Railway : The British Experimental Line in Asia Minor

Yakup Bektas

Introduction

The establishment of railways in Asia Minor was at the outset a concern primarily of the British. They saw a system of rail communication in that region as an essential part of their eastward imperial expansion. With the close of the Crimean War, the construction of such railways came to offer not only greater commercial potentials, but also increasing political and strategic advantages, especially in linking India, Asia and Trans-Caucasia to Britain. While public and official debates continued in Britain over the Euphrates Railway — a grand Victorian railway project to connect Europe with India and Central Asia via the Euphrates valley railway — British entrepreneurs initiated plans building relatively short railways to link the regions of rich agricultural production and valuable raw materials in the interior of Asian Minor with seaports, thus allowing greater access to such agricultural and raw goods for overseas trade. Most of these small rail lines were originally proposed as profit-oriented ventures rather than major political or imperial projects, and investors and builders made certain that the railway projects would return their money many times over. The projects thus proposed by British and European companies included : on the Aegean Sea, the Izmir-to-Aydin and Izmir-to-Kasaba (Turgutlu) lines ; on the Sea of Marmara, the Mudanya-to-Bursa line and lines terminating at Üsküdar (Haydarpasha) and Izmit ; on the Black Sea, lines terminating at Trabzon, Samsun, Sinop and Costanza ; and on the Mediterranean, lines terminating at Mersin and Adana[1].

In this paper I will examine the British efforts to " pioneer " the construction of railways in the Ottoman Empire as reflected in the case of the Izmir

1. On the railways in the Ottoman Empire proposed by European capitalists, see C.E. Austin, *Underdeveloped Resources of Turkey in Asia, With Notes on the Railway to India,* London, 1878, 61-74.

(Smyrna)-to-Aydin Railway, the first railway line in Asia Minor, commenced in 1857. The builders of the railway presented it as an " experimental line " and a model commercial enterprise in the Ottoman Empire. Later the perceived success (or lack thereof) of the project would be invoked as an example in the planning of other schemes. The railway was constructed exclusively by British engineers and management, and only British materials were used.

Within two decades after the onset of " railway mania " in Britain in 1830, when steam locomotive trains began to run, railways had become the ultimate symbol of European industrial and economic power. The success of the railways precipitated a surge of subsequent industrial and managerial innovations. Railways thus came to be seen the key to modernization, progress and economic growth. Important, as well, was the tremendous and immediate visual impact of the railways, and the public image that they generated for the society that possessed them[2].

By the mid-nineteenth century, after a period of rapid construction, the major network of railways in Britain was largely complete[3]. British railway engineers and investors then turned their attention abroad — first to Western Europe, where the completion of major rail networks was already well underway, and then to the U.S.A., where railways were growing rapidly, and on a grand scale. Next came the non-Western world. As early as 1850, Britain and France had begun to envision and construct new railways outside of their territories, all over the world. These railway projects were often undertaken in connection with broader imperialist objectives[4].

The Imperial Medjidieh-Ottoman Railway

Sir William Fairbairn, the celebrated British engineer, arrived in Istanbul in 1839 as a part of a " scientific mission " organized according to a directive of Sultan Mahmud II[5]. The Sultan had earlier dispatched a commission to Britain and other Western countries, for the purpose of making enquiries about " useful arts and manufactures " that might be introduced to his country, and to seek out men of " practical science " to survey and report on the different

2. For a study of the image and impact of railways, see N. Faith, *The World the Railways Made,* London, 1990.

3. H. Pollins, *Britain's Railways*, Newton Abbot, 1971.

4. See D.R. Headrick, *The Tentacles of Progress : Technology Transfer in the Age of Imperialism*, 1850-1940, Oxford 1988 ; *The Tools of the Empire : Technology and European Imperialism in the Nineteenth Century,* Oxford, 1981.

5. Fairbairn appears to have arrived in Istanbul a few days before the death of Sultan Mahmud II. In the same letter Fairbairn writes that his majesty died on the very morning on which he was to have had his audience at the palace. Due to the death of the Sultan, Fairbairn's inspections and surveys of the public works were temporarily suspended until he had received new orders from the Grand Vizier to proceed. See *The Life of Sir William Fairbairn* (partially written by Fairbairn himself, edited and completed by William Pole). First published in 1877 by Longmans Green & Company. I consulted the 1970 reprint, especially Chapter XI, 165-176.

establishments then in operation in the Ottoman Empire[6]. Fairbairn was one of these men of " practical science ", brought in to inspect and advise on industrial and technical operations in the Ottoman Empire[7]. He made valuable observations on the state of public works in Turkey, and his remarks on the Ottoman industrial enterprises included railways. His long meetings with high officials, " pashas and effendis " connected with various government departments (such as War and Ordnance) were focused on " the improved state of practical science in England ", and the potential introduction of railways into the Ottoman Empire. Fairbairn reported of these meetings : " (Railways) appeared to them inexplicable, if not entirely beyond their comprehension. They could not realise the idea of travelling at the rate of forty miles an hour, and doubted the correctness of the descriptions that reached them "[8].

At that time railways had not yet been extended beyond Western Europe. It was only during the Crimean War (1853-1856) that the Ottomans first experienced a railway system. During that war, the British built a wartime railway, surveyed and engineered by D. Campbell, between the camp of the Allies at Sebastopol and their base of supplies at Balaklava, about eight miles away. The railway was very simply constructed : rails were fastened down over wooden sleepers which were laid over a bed of stones on the road[9]. It is the earliest instance of a purely military railway being constructed during a war[10].

The Sultan and his officials developed a strong interest in rail construction once the practicality of the railroad had been demonstrated to them in the Crimean War. In late 1855, even before the end of the war, the Ottoman government declared and transmitted to the various Embassies and Legations throughout Europe " the fundamental conditions " for the establishment of railways in the Ottoman Empire. To this end the government " resolved to address itself directly to the experiences and capital of Europe "[11]. It determined that the construction and operation of Ottoman railways should be under the high

6. *Ibid.*, 166.

7. Fairbairn acknowledges that he reported on nearly all the Ottoman government works. Later, on his recommendation, an Ottoman commission under Mr. Ohannes Dadian (Assoc. Inst. CE) visited England, " in furtherance of the plans for ameliorating the state of the Turkish community by introducing useful arts and manufactures ". See *Minutes of Proceedings of the Institution of Civil Engineers,* II (1848), 125-6 ; William Fairbairn, " Experimental Researches into the Properties of the Iron Ores of Samakoff, in Turkey, and of the Hematite Ores of Cumberland, with a View to Determine the Best Means for Reducing them into the Cast and Malleable States ; and on the Relative Strength and Other Properties of Cast-Iron from the Turkish and Other Hematite Ores ", *Ibid.*, III (1844), 225-241.

8. Fairbairn, *Life of Fairbairn*, 170-171.

9. Brian Cooke, *The Grand Crimean Central Railway, the Railway that Won a War : The Story of the Railway built by the British in the Crimea during the War of 1854-1856*, Cheshire, 1990.

10. E.A. Pratt, *The Rise of Rail-Power in War and Conquest,1833-1914,* London, 1916, 206-211.

11. *The Times*, 2nd October, 1855, 8-9.

superitendence of the Sublime Porte. A company to construct a line could be established only under the name and title of *The Imperial Ottoman Company.* This company would at all times be subject to the general laws of the Ottoman Empire and remain its property. Foreigners were granted the right to act as shareholders, without any distinction from shareholders who were Ottoman subjects.

After the war, the Europeans were taken by surprise by how quickly the Sultan and his government passed *a firman,* or Imperial Charter, allowing European companies to build railways in various parts of the Ottoman Empire. Referring to the concession granted for the construction of a railway linking the Danube to the Archipelago port, a *Times* correspondent remarked that " there was never a project presented to the Ottoman government which was so universally approved by all or found fewer objections "[12]. A privately circulated official memorandum of the Ottoman Tanzimat Administration declared : " To develop to the utmost extent the resources of the Empire, to bring its capital and its most productive provinces into the shortest communication with the capital and with western Europe, and at the same time to complete the European portion of the direct route to India, it is proposed to construct the *Imperial Ottoman Railway* "[13].

As evidence of his personal appreciation of this undertaking, the Sultan wished his own name should be connected with it, and consequently that it should be the *Imperial Medjidieh-Ottoman* Railway[14]. The line was to connect the Danube at Rustchuk to Varna and Istanbul, then connect these via rail links with the Black Sea and the Mediterranean. This line was expected to form " the great high road between the East and the West ", and to be the " main artery " of the railway system of the European portion of the Ottoman Empire, providing an outlet for the vast produce of the richest, best inhabited and most productive districts of Ottoman Europe. However, these ambitious projections remained largely that, as did the Euphrates Valley Railway as well.

Although Ottoman officials were aware of the military and economic benefits of a rail network in their empire, the country lacked technical and entrepreneurial infrastructure, as well as the finances to build any such infrastructure. Therefore, Europeans were invited to construct and operate railways in the Ottoman Empire, where, with the close of the Crimean War, a large number of government grants and European loans had been made available for railway enterprises. Having no experience in business transactions with Europe, the Ottoman government relied heavily on the mediation of opportunist Levan-

12. *The Times*, 30th January, 1857.

13. *The proposed Imperial (Medjidieh) Ottoman Railway, its Purposes and Prospects.* Printed for private circulation only, 1857, London ? or Constantinople ?, British Library, Miscellaneous Tracts, 82925-1915.

14. *Ibid.*, 4.

tines[15], who were more in touch with Europe and European languages.

Offering their services mainly as *dragoman* (translators), agents and intermediaries, the opportunist Levantines took advantage of Ottoman ignorance of European business and entrepreneurial transactions for their own economic benefit. They got involved in the railway concessions granted by the Ottoman government to European entrepreneurs as agents. However, they turned the " concessions " to a profitable business[16]. Referred to as a " gang of speculators " and " half-caste adventurers " by *The Engineer,* they indeed had no genuine knowledge of railway construction and enterprises[17].

They used every means — even bribery or pretending to be the representative of some European investor — to gain concessions from naive pashas. The concessions then would be sold to European investors, who were often ignorant of the country in which they were operating. In this way, the opportunist Levantines made large sums of money. But there were too many concessions sold in this way for projects that could not practically be realized. The complexity of sorting this out deterred the leading engineers and contractors from engaging in Ottoman railways. After some examination, most projects advertised by opportunist Levantines would be found impracticable[18]. Thus only very few of the projected rail lines were constructed before 1860. One of these was the line from Izmir to Aydin.

THE OTTOMAN RAILWAY FROM IZMIR TO AYDIN OF HIS IMPERIAL MAJESTY THE SULTAN

Situated on the Aegean Sea near the eastern extremity of the Mediterranean, Izmir (Smyrna) was a major commercial port city in Asia Minor[19]. The city had grown up on lands adjoining two major rivers. The railway would enable Europeans to transport their manufactured goods from the sea port at Izmir to towns deep inside Asia Minor. In return, raw materials could be carried back to Izmir, and from there to European ports. In the mid-nineteenth century a special commissioner estimated that the commercial traffic between Aydin and Izmir employed 10,000 camels and 500 mules at a cost of over 400,000 pounds sterling per annum. Independent of these camels, which were constantly

15. In this paper, I use the term " Levantine " as an actors' category to describe a group of men who acted as middlemen or brokers between European entrepreneurs and Ottomans. They belonged to various nations, including the Ottoman Empire and its diverse minorities, and were somewhat familiar with European languages and enterprises.

16. H. Clarke, " Railways in Turkey I ", *The Engineer,* 21st September, 1867, 344.

17. *Ibid.*

18. *Ibid.*

19. For a study of the economic and commercial significance of Smyrna at the time, see Mr. F. Wakefield, " Report on Smyrna and its Producing Districts ", 1857, in Sir R.M. Stephenson, *Railways in Turkey : Remarks upon the Practicality and Advantage of Railway Communication in European and Asiatic Turkey*, London, 1859, 30-38.

employed to carry goods between the two towns, there were at least another 20,000 camels employed along various routes for conveying produce from the interior to the sea[20]. The Austrian Consul at Smyrna, C. Zalloni's trade charts of Smyrna for 1857 shows imports for that year amounting to 2,447,493 pounds sterling and exports amounting to 2,397,342 pounds sterling[21]. The terms of this report illustrate that economic interests were preponderant in this undertaking.

The concession for the Izmir-Aydin railway was originally granted to a British group consisting of Joseph Paxton, George Whites, William Jackson, and A. W. Rixon[22]. The *firman* issued to them declared the concession a free grant by the Imperial Ottoman government to the concessionaires to encourage the completion of the proposed rail line[23]. However, in the company's budget the concession appeared to have cost them 40,000 pounds[24]. This money must have been shared among the company's agents or the opportunist Levantines. The contract was originally for 72 miles of railway plus harbor works, and stock was taken by Thomas Jackson at 1,030,000 pounds[25]. The concession was granted for 50 years beginning in 1860, by which time the rail line was expected to be ready for commercial traffic. The Ottoman government guaranteed the company a net income of six percent on the capital invested in the actual construction of the line. This capital expenditure was not to exceed 1,200,000 pounds[26].

The capital was to be obtained by an issue of shares, with one fourth of the shares being reserved for the Ottoman Empire. Subscription was accordingly invited in 1857-8 for 60,000 shares at 20 pounds each. By the time work began on the railway, 49,478 of the total shares were subscribed : 15,000 by the Ottoman government, 500 by the Sultan, 25,750 by the contractor, 6,500 by the concessionaires, and 1,728 by the British public[27]. The company was established in London with the name of *The Ottoman Railway from Smyrna to Aidin of His Imperial Majesty the Sultan.*

20. R.M. Stephenson, *Railways in Turkey*, 6 ; F. Wakefield, " Report on Smyrna ", 30-37.

21. " Abstract of Import and Export Trade of Smyrna from Mr. Zalloni's Tabes for 1857 " in R.M. Stephenson, *Railways in Turkey*, 46-48.

22. For the original convention reached on 23rd September, 1856, see " Izmir-Aydin Demiryolu Imtiyaz Mukavelenamesi ", " 23 Eylul, 1856, Osmanli Mevzuati Hukukiyesi ", *Sicill-i Kavanin : Osmanli Mevzuat-i Hukukiyesi, 23 Eylül 1856 I, 5 Mart 1917 II* (1935 edition), 1-11 ; *Mecmua-i Mukavelat,* I (1856), 2.

23. *Ibid.*

24. H. Clarke, " Railways in Turkey I ", *The Engineer*, 344.

25. *Ibid.*

26. *Ibid.*

27. " Memorandum on Smyrna and Aidin Railway Company by the Committee of Investigation ", FO (Foreign Offce) 78/2255, *Smyrna and Aidin Railway (Ottoman Railway), 1867 to 1872.* PRO (Public Record Office), London.

Rowland MacDonald Stephenson, a celebrated civil engineer and contractor and the managing director of the East Indian Railway Company, was made the chairman of the company[28]. A man of grand Victorian projects, Stephenson planned railways between Europe, India and Central Asia[29]. As the first work of its kind in the Ottoman Empire, the successful completion of the Izmir-Aydin railway would have a special impact on all similar future enterprises.

The work began in mid-autumn of 1857. The tracks were laid on the British gauge, 4 feet 8.5 inches. When work started, the Chief Engineer was George Meredith, who had been involved in many railway projects in Ireland, Holland and Britain, including the construction of the Liverpool and Manchester Railway under George Stephenson[30]. The commencement of the railway works in Izmir was a big occasion. *The Illustrated London News* remarked : " It appeared as if the whole city of Smyrna was formed in an animated circle around the field where a ceremony celebrated this great undertaking, which will establish *a new era* in Turkey "[31].

Indeed, Izmir was a well chosen port city not only for its location, but also on account of its demographic, social and cultural setting. Its population consisted of many different ethnic groups, including Turks, Greeks, Armenians, Jews, and many Europeans. Thus it was a city that accommodated different religions, languages and traditions.

The locomotive

The occasion of the introduction of the first locomotive in the Ottoman Empire was used by the company to display their technological power. A locomotive was brought in when not even ten miles of the line had been completed. It was an impressive show for the public and the local administration, who were thought to be suspicious of the railroad. *The Times* remarked : " They heard that if that Frankish *marifet* (marvel) was finished they could go in two hours from Aidin to Smyrna, a journey which, with considerable exertion, they

28. Sir Rowland MacDonald Stephenson (1808-1896), born in London and educated at Harrow, became a civil engineer in 1830. He was secretary to the association established in 1835 for securing steam communication with India, and managing director and also deputy chairman of East Indian Company. He was the author of some scientific books on railways, such as *Science and Railway Construction* (4th ed. by Nuget,1869). See *ILL*, 30 (1857), 338 ; *Minutes of Proceedings of Institutions of Civil Engineers*, 128 (1896), 451-62.

29. Later Stephenson was involved in a project to construct the first railway in China, in 1864, where upon the suggestion of Jardine and Matheson & Co., he undertook a preliminary survey for an " experimental line " between Shanghai and Woosung. See Alan Reid, *The Woosung Road : the Story of the first Railway in China 1875-1877*, Suffolk, 1977, 1-14.

30. In addition to his service for the Ottoman Railway Company, he made a survey of a proposed line from Istanbul to Edirne and returned to Britain with the intention of taking measures to have this line carried built, but he died soon afterwards. *Minutes of Proceedings of Institution of Civil Engineers,* 25 (Session 1865-66), 515-6 ; *The Times*, 16th November,1858.

31. *The Illustrated London News,* 436, 31st October,1857.

may now accomplish in two days ; they shook their heads and asked whether it was by flying through the air. Those who lived in Smyrna, and those who visited it, looked with astonishment at the mounds of earth thrown to the right and left, at the rocks which were blasted and forced out, and the mysterious iron lines which were drawn with cabalistic calculations, without being able to understand how all these contrivances could in any way advance the object in view "[32].

From the beginning, the railway works had been the center of attention in Izmir. Bearing in mind the general suspicion the local public felt toward European innovations and machinery, the builders expected the first locomotive to leave a powerful visual impression on the Ottoman Empire. To the *Times* correspondent, who was witnessing the arrival of the first locomotive in Asia Minor, the point was to " make an epoch in the minds of the people in this part of the world "[33].

Curiosity was at its peak when " a black-looking vessel was seen to drop its anchor in the bay to the west of town, fronting the line. It contained the mysterious engine "[34].

It was met by crowds of people at Smyrna. They were described as belonging to not less than nineteen different races and nationalities — English, Irish, Scotch, French, Americans, Italians, Slovenians, Armenians, Turks, Greeks, Poles, Albanians, Austrians, Prussians, Hindus, Africans, Ionians, and Spaniards[35].

The locomotive, named St. Sophia and decorated with both the Ottoman standard and the Union Jack, carried forth a band of musicians, who played the " Sultan's March ". It first passed a scarcely-finished building where the owner, a local Turk, pointed up the words " Railway Hotel ". He asked for a drawing representing a locomotive as his sign board[36].

The company and its backers were convinced that the running of a locomotive, however premature, would do more than anything to convince the Ottoman public of the serious nature of the enterprise and erase any last doubts. Stratford de Redcliffe, the British Ambassador at Istanbul, who went to Izmir on the occasion of the laying of the foundation stone of the Izmir Station in November 1858, firmly believed that the success of this first railway would be a foretaste of similar triumphs in other parts of the vast Ottoman Empire. In the future, this line would be one leg in a great network of iron communication stretching across Asia Minor. De Redcliffe strongly advocated his belief that

32. *The Times*, 6th April,1858.
33. *Ibid.*
34. *Ibid.*
35. *Ibid.*
36. *ILL*, 27th November, 1858, 512-3.

the railway and telegraph were the basic means of Western civilization. By acquiring these means, the Ottoman Empire was expected to become a part of Western civilization[37].

With the introduction of the first locomotive, the need to systematize the construction of railways in Turkey under certain laws and guidelines became a main concern both of the builders and of the Ottoman government. When he visited Izmir in 1858 to inspect the progress of the works, R.M. Stephenson was invited by the Ottoman ministers to advise on the establishment of a wider network of railways in the Ottoman Empire. Stephenson then chaired the meetings of the Constantinople Committee on Railways in Turkey. The Committee resolved to set down basic rules and conditions before embarking on future railway enterprises. The most important of these rules concerned the surveying of the country, the terms of future concessions, the setting of tariffs, the generation of by-laws, and the regulation of the companies[38]. After the meetings, Stephenson made a report to the Grand Vizier, Ali Pasha, for the consideration of the Porte[39]. Meanwhile, to encourage the regulated development of the Ottoman railways, the government set up a Department of Public Works, with special responsibility for railways and telegraph lines. This department was responsible for keeping a *cahier de charges,* or general code of regulations under which future lines in the Ottoman Empire were to be sanctioned[40].

" Waste, robbery and peculation "

During the first year of its construction, the work on the Izmir-to-Aydin Railway advanced satisfactorily. R. M. Stephenson reported that the line employed about 3,000 men of seventeen different nations[41]. A 10-mile section of track from Izmir to Seydiköy was opened on October 30, 1858. The opening was carried off in an impressive display intended to win the confidence of the local administration and residents.

However, after the first year the construction of the line did not progress as planned. By the end of 1860, the date originally proposed for the completion of the whole line, only 27 miles of track had been built, bringing the line from Izmir to the Tiranda Station. Meanwhile, at Aydin, the company had only just

37. See the extracts from his speech on the occasion of the laying the foundation stone of Smyrna Station and opening of the first section of line. R.M. Stephenson, *Railways in Turkey,* 37-46.

38. See " Minutes of Proceedings of the Constantinople Committee upon Railways in Turkey ", R.M. Stephenson, *Railways in Turkey,* 53-56.

39. See " Copy of Letters Addressed to the Ministers at Constantinople ", in R.M. Stephenson, *Railways in Turkey*, 49-52.

40. See " Chair des Charges de la Concession d'un Chemin de Fer " in R.M. Stephenson, *Railways in Turkey,* 63-80.

41. R.M. Stephenson, *Railways in Turkey*, 3-4.

begun work on the tunnel through Saladdin Pass[42]. Soon after, the engineers suggested an alternative route, and the Saladdin tunnel was abandoned. The line was built instead through the Ephesus Pass. This deviation from the originally planned route lengthened the line by about ten miles, making the total length about 81.5 miles.

The need to revise the planned route in mid-construction suggests that the engineering survey had not been done properly at the outset. But the problems were not limited to engineering difficulties : In addition, the Ottoman government placed part of the blame for the slow progress on the suspension of construction work by the contractor[43]. The company went into bankruptcy and its shares became unmarketable. The company's Chief Engineer, Second Engineer, Secretary, General Manager, and agents in both Istanbul and Izmir were removed without suitable replacements being found by the Chairman — who himself disappeared for a while[44]. Hyde Clark summed up the general feeling back in Istanbul in *The Engineer* : " The Smyrna-Aydin Railway was nothing but waste, robbery, and peculation, administered by rouges and swindlers, with the complicity and connivance of fools and knaves in England "[45].

Not surprisingly, in light of all this, the cost of construction was greater than anticipated, and the original limit on capital expenditures of 1.2 million pounds proved too low. By this time, the state of the railway enterprise was creating tension in both the Ottoman Empire and Britain. Moreoever, the problems plaguing the Izmir-Aydin line stood in the way of the construction of other lines proposed by new promoters. In March, 1861, the government agreed to grant an extention of three years on Izmir-Aydin construction, until May, 1864. The company now had a new contractor, T.R. Crampton. Other important new officials included Edwin Clark, Consulting Engineer, Edward Purser, Chief Engineer, S.J. Cooke, Secretary, and W.F. Fergusson in charge of Traffic and Locomotive Support[46]. By November, 1861, 41 miles of track stretching from Izmir to Karapinar had been opened. However, the full line reaching all the way to Aydin was only completed in July, 1866, after yet further delay. The station buildings were completed only after the opening of the full line.

42. " Official Communication " of the Board of Public Works in response to the " Memorandum of the Three Anglo-Turkish Railway Companies to the Turkish Government " (dated 29th April,1868), September, 1868, F078-2255, *Smyrna and Aidin Railway (Ottoman Railway),1867 to 1872.*

43. *Ibid.*

44. *Ibid.* ; see also " Memorandum on Smyrna and Aidin Railway " by the Ottoman Railway Company, F078-2255, *Smyrna and Aidin Railway, 1867 to 1872*, 225-228.

45. H. Clarke, " Railways in Turkey II ", *The Engineer*, 1st November, 1867, 372.

46. " Official Communication " of the Board of Public Works in response to the " Memorandum of the Three Anglo-Turkish Railway Companies to the Turkish Government " (dated 29th April, 1868), September, 1868, F078-2255, *Smyrna and Aidin Railway (Ottoman Railway), 1867 to 1872.*

THE IZMIR-KASABA RAILWAY

The Izmir-Aydin railway had been intended to provide a model for railway enterprises in Asia Minor. However, in the end it did not make for a very impressive example. On the contrary, the railway's long-delayed arrival and the mismanagement during the construction process only served to increase the frustration of local officials and the general public with the company. On top of all this, the overall cost of the railway turned out to be much higher than anticipated. Hoping to end the monopoly of the Izmir-Aydin Company within a decade, the government sanctioned the establishment new British and European companies.

In spite of the Izmir-Aydin Company's opposition to the granting of concessions to any other company, in 1863 the Ottoman government granted a new concession to Edward Price, a British railway engineer and contractor who had a link with Robert Stephenson. Price proposed to build a new line from Izmir to yet another town in the interior, Kasaba (now Turgutlu)[47]. He soon transferred the concession to the British company which had been set up in London to construct the line, named (in the British redaction) *The Smyrna-Cassaba Railway*. The company began construction in 1864, and had completed the largest section, from Izmir to Manisa (41 miles), by October, 1865. The remaining 17.25 miles to Kasaba was completed in the following year, and the whole line was opened to traffic even before the Izmir-Aydin line in 1866 ! The company received the praise of the Ottoman government : " We have an example in the railway from Izmir to Kasaba, which has been constructed with economy and administered with intelligence. The revenue generated by this railway very nearly matches the guarantee from the state, and the company in question has never had any dispute with the state, nor any trouble. This shows that in Ottoman Turkey, just like elsewhere, railways can succeed, provided they are in the hands of administrators who are clever, economical, and above all, sincerely loyal in their duties to the public and the state "[48].

In contrast with the Izmir-Aydin Company, the Izmir-Kasaba Company encountered no important management or construction difficulties. It obtained another concession in 1871 for an extension of the Izmir-Kasaba line to Alashehir, a further 47.50 miles inland. Like the Izmir-Aydin line, both the original Izmir-Kasaba line and the Alashehir extension were built on the British system, using broad gauge. All orders for engines, rolling stock, and other railway materials were placed with British companies. Locomotives for the

47. For Edward Price see J. Marshall, *A Biographical Dictionary of Railway Engineers*, David & Charles, 1978, 174 ; also *Min. Proc. ICE*, 33 (1871-2), 267-269.

48. " Official Communication " of the Board of Public Works in Response to the " Memorandum of the Three Anglo-Turkish Railway Companies to the Turkish Government " (29th April, 1868), September, 1868, F078-2255, *Smyrna and Aidin Railway (Ottoman Railway), 1867 to 1867.*

Izmir-Aydin Railway were purchased from Robert Stephenson Co. Locomotives for the Izmir-Kasaba Railway were purchased from Beyer Peacock, 4-4-0 and 0-4-2 types[49].

In the summer of 1867 Sultan Abdulaziz visited several European countries, including Britain. He was the first Ottoman Sultan ever to visit Europe. While on this trip, the Sultan attended the Paris Exhibition of 1867. It was also on this trip that the Sultan made his first railway journey, beginning in his own territory, in Rutschuk. The line from Rutschuk to Varna, built and operated by the British, was about 120 miles in length, and ran through some of the finest hill and forest landscapes of Bulgaria, then still an Ottoman province.

The Engineer (1867) declared that " the Sultan's visit brought Turkey nearer to Europe "[50]. While in Europe, the Sultan expressed his desire to extend railways throughout the Empire[51]. The trip proved an excellent opportunity for the Sultan to experience railways and other industrial accomplishments. The *Times* reported : " Of all the strange things the Sultan saw on his journey, nothing seems to have produced a greater impression than the rapid means of locomotion by rail ; nothing seems to have caused greater satisfaction than re-entering his own dominions by the Rutschuk and Varna Railway, a railway in his own dominions "[52].

It was reported that after the Sultan's return the two main subjects to which state attention was turned were communication and Western-style schools. There were a considerable number of new negotiations with entrepreneurs seeking to build railways in the Empire. A prominent French railway engineer, J. H. Haddan, acknowledged in 1873 that Istanbul was " full of engineers "[53]. Sultan Abdulaziz was now keen to see an extensive network of railways intersecting across the whole of the Empire. Referring to the political disturbances resulting from this obsession of the Sultan's, Count N. P. Ignatyev, the Russian ambassador at the time, reported that the Sultan was " victim of a veritable railroad fever "[54].

In the early 1870s the British system of rail construction came under some criticism, especially by the French. Haddan denounced the British broad-gauge system on the grounds that it was an extravagant use of materials, and not well-suited to either the geographical and financial conditions of the Ottoman

49. J.G.H. Warren, *A Century of Locomotive Building by Robert Stephenson & Co., 1823-1923*, Newcastle, 1923, 612-614 ; E. Talbot, *Steam in Turkey : An Enthusiasts' Guide to the Steam Locomotives of Turkey,* London, 1981, 41.

50. H. Clarke, " Railways in Turkey III ", *The Engineer,* 22nd November, 1867, 436-437.

51. *The Railway News*, 21st September, 1867, 296-297.

52. *The Times*, 15th January, 1868, and 28th January, 1868.

53. J.L. Haddan, *La Largeur de Voie Convenable pour Les Chemins de Fer de la Turquie,* Imprimerie et Lithographie Centrales, Constantinople 1873, 8. I thank Ana Carneiro for this reference.

54. R.H. Davison, *Reform in the Ottoman Empire, 1856-1876*, Princeton, 1963, 237.

Empire[55]. Haddan believed that the use of narrow gauge rails, only 3 feet wide, would reduce the cost of construction by well over 50 percent. At an Ottoman scientific *madjlis*, or convention, Haddan urged those concerned to re-examine the problems relating to the gauge of the railways[56]. However, a committee to examine the issue was never established, due to ongoing political instability and the general lack of entrepreneurial experience.

CONCLUSION

After the Crimean War, railways began to attract an increasing attention from the Porte — and from the Sultans personally. This interest on the part of the Ottoman state became particularly intense in the late 1860s, following Sultan Abdulaziz's visit to Europe. Railways represented a new image of a powerful and industrially developed Europe. After opening to goods and passenger traffic, the Izmir-Aydin and Izmir-Kasaba railways enjoyed a fair measure of commercial success. This further stimulated the railway craze within the Ottoman government, as well as wider public interest.

The Sultan and his government's ambition to extend railways throughout the empire was hampered by a lack of entrepreneurial skills, political stability, and financial power. The industrial-technological infrastructure required for railway building could not be adequately established. Thus the Ottoman railways were from the beginning financed, constructed, and operated by European entrepreneurs. The first railway project by the British, the Izmir-Aydin line, a relatively short line whose completion had been expected to take three years, was completed only after a decade of work. All along the way, the opportunist Levantines and European railway speculators exploited Ottoman ignorance of European-style entrepreneurship to their own advantage. It is only during the long reign of Sultan Abdulhamid II (1876-1908) that railways became politically and militarily powerful within the Ottoman Empire. At that time, multinational companies were encouraged to build railways, and Germany emerged as the strongest player, while Britain lost new concessions. Indeed, the history of Ottoman railway-building in the late nineteenth century presents an image of European powers competing for a bigger share in the Ottoman Empire itself. With the final decline of the Ottoman Empire, railways increasingly became projects more political than commercial in character.

Throughout, the Ottomans remained almost entirely dependent on European hardware, knowledge, and technical skills. In constrast to electric telegraphy, which had been extended throughout the empire within just over a decade of its introduction, rail construction involved a larger technological system set up

55. J.L. Haddan, *La Largeur...*, 1-8.
56. *Ibid.*, 7-17.

at a far higher cost, and required heavy industrial components and sophisticated engineering skills.

ACKNOWLEDGEMENTS

I wish to thank Tadaaki Kimoto for his kind help and advice, Crosbie Smith for his guidance during the initial writing of this paper, and Ana Carneiro, Elizabeth Pidoux, Ben Marsden, Jon Agar, and Philip Chaston for their encouragement. I gratefully acknowledge the financial support of the Japan Society for the Promotion of Science.

FIGURE

The commencement of the Izmir and Aydin Railway, gathering Ottoman ministers, high ranks pashas, and civil and military authorities of Izmir ; Mufti (High Priest), the Mallah (Judge), the Imams (priests), the Consuls, the Greek and Armenian Bishops, the Great Rabbi of the Jews, and the general public. The railway raised the interest of all ethnic and religious communities, and occasioned a forum for alliance and sodality. (From *The Illustrated London News*, 31st October 1857, 436).

Contributors

Mohamed ABALLAGH
Université de Fès
Fès (Maroc)

Yakup BEKTAS
Tokyo Institute of Technology
Tokyo (Japan)

Patrice BRET
Cité des Sciences et de l'Industrie
Paris (France)

Jean DHOMBRES
Ecole des Hautes Etudes
en Sciences Sociales
Paris (France)

Ahmed DJEBBAR
Université de Paris-Sud
Paris (France)

Youcef GUERGOUR
Ecole Normale Supérieure
Alger (Algérie)

Feza GÜNERGUN
Istanbul University
Istanbul (Turkey)

Ekmeleddin IHSANOGLU
Istanbul University
Istanbul (Turkey)

Mustafa KAÇAR
Istanbul University
Istanbul (Turkey)

A.H. Helmy MOHAMMAD
Ain Shams University
Cairo (Egypt)